MATHEMATICAL DISQUISITIONS

MATHEMATICAL DISQUISITIONS

The Booklet of Theses
Immortalized by Galileo

* * * * *

CHRISTOPHER M. GRANEY

University of Notre Dame Press

Notre Dame, Indiana

University of Notre Dame Press
Notre Dame, Indiana 46556
www.undpress.nd.edu

All images from the *Mathematical Disquisitions* are used
courtesy of ETH-Bibliothek Zürich, Alte und Seltene Drucke.

Published in the United States of America

Library of Congress Cataloging-in-Publication Data

Names: Graney, Christopher M., 1966–
Title: Mathematical disquisitions : the booklet of theses immortalized
by Galileo / Christopher M. Graney.
Description: Notre Dame, Indiana : University of Notre Dame Press, [2017] |
Includes bibliographical references and index. |
Identifiers: LCCN 2017030443 (print) | LCCN 2017038797 (ebook) |
ISBN 9780268102432 (pdf) | ISBN 9780268102449 (epub) |
ISBN 9780268102418 (hardcover : alk. paper) | ISBN 0268102414
(hardcover : alk. paper) | ISBN 9780268102425 (pbk. : alk. paper) |
ISBN 0268102422 (pbk. : alk. paper)
Subjects: LCSH: Locher, Johann Georg. Disquisitiones mathematicae. English |
Galilei, Galileo, 1564–1642. Dialogo dei massimi sistemi. English. |
Astronomy—Early works to 1800. | Sun—Early works to 1800. |
Sunspots—Early works to 1800.
Classification: LCC QB47 (ebook) | LCC QB47 .G73 2017 (print) |
DDC 520—dc23
LC record available at https://lccn.loc.gov/2017030443

∞ *This paper meets the requirements of ANSI/NISO Z39.48-1992
(Permanence of Paper).*

To my sister,

Laura Kathleen Graney (1971–2013).

She liked the moon, and my work.

She would have liked Disquisitions 25 through 27,

and how no one had read them for a long time.

CONTENTS

Mathematical Disquisitions,
Concerning Astronomical Controversies and Novelties

ACKNOWLEDGMENTS

Dennis Danielson first directed my attention to Johann Georg Locher and his *Disquisitiones mathematicae.* Matt Dowd of the University of Notre Dame Press expressed early interest in a translation of Locher's book, shepherded this project along, and introduced me to Darin Hayton of Haverford College. Darin served as a valuable consultant and referee on the translation. Absent Dennis, Matt, or Darin, this book would not exist in anything like its present form. I owe them all many thanks. I also owe many thanks to my wife, Christina Graney. She walked through the entire translation with me, insisting that I not drift too far from the original Latin in my effort to produce a student-friendly book. The idea for a student-friendly translation is entirely mine. The responsibility for the flaws that exist in it is likewise entirely mine.

I thank E-rara and Google Books for making high-resolution copies of Locher's original work freely available on the Internet. I thank the Louisville (Kentucky) Free Public Library, whose resources I used extensively in this project. I also thank my college, Jefferson Community & Technical College in Louisville—it is the academic soil in which I have grown.

Introduction

We know Johann Georg Locher because Galileo Galilei immortalized him as an exemplar of anti-Copernican silliness. Without Galileo, Locher might have vanished into obscurity.

But Galileo devoted many pages of his 1632 *Dialogue Concerning the Two Chief World Systems: Ptolemaic and Copernican* to Locher's short book of 1614 entitled *Mathematical Disquisitions Concerning Astronomical Controversies and Novelties*. This is the "booklet of theses, which is full of novelties"[1] that Galileo has his less-than-brilliant character Simplicio drag out in order to defend one or another wrongheaded idea. When Galileo (through his character of Salviati) describes the author of this booklet as producing arguments full of "falsehoods and fallacies and contradictions,"[2] as "thinking up, one by one, things that would be required to serve his purposes, instead of adjusting his purposes step by step to things as they are,"[3] and as being excessively bold and self-confident, "setting himself up to refute another's doctrine while remaining ignorant of the basic foundations upon which the greatest and most important parts of the whole structure are supported,"[4] he is speaking of Locher. He is also defining Locher (and anti-Copernicans in general), especially for modern readers who study the debate over the heliocentric world system of Nicolaus Copernicus by means of translations of the *Dialogue* or of Copernicus's 1543 *On the Revolutions*. And Galileo is not defining Locher alone.

Disquisitions has always been largely credited to Locher's mentor, the Jesuit astronomer Christoph Scheiner, under whose supervision it was published.[5] Galileo also devotes pages of the *Dialogue* to discussing Scheiner's work on sunspots.[6] Thus the *Dialogue* pertains all the more to the work of, and to defining, Locher and Scheiner. Indeed, one of the consultants asked by the Inquisition to study the *Dialogue* for Galileo's trial in 1633 described Galileo's principal aim within the book as attacking Scheiner.[7] Galileo immortalized Locher and Scheiner through criticism of them.

Modern readers may therefore be surprised to find that even leafing through Locher's *Disquisitions* raises questions regarding Galileo's portrayal of anti-Copernican thinking (Figure I-1). For example, in the *Dialogue* Simplicio argues, based on Aristotelian ideas about the heavens, for a moon that is smooth. He says that those things seen on the moon through a telescope, "mountains, rocks, ridges, valleys, etc." are "all illusions."[8] But *Disquisitions* contains a page-width illustration of the moon, showing these supposed illusions in detail. The *Dialogue* portrays the two chief world systems as being "Ptolemaic and Copernican," but leafing through *Disquisitions* reveals that the two systems most carefully illustrated within it (in detailed full-page diagrams) are the Copernican system on one hand and the hybrid geocentric system of Danish astronomer Tycho Brahe (in which the sun circles Earth while the planets circle the sun) on the other. *Disquisitions* also contains an illustration of the sun with spots, an illustration of Venus showing phases as it circles the sun, and two remarkable pages of illustrations of the Jovian system. One of these pages contains a diagram of the system complete with the orbits of moons, the Jovian shadow, indications of the points where eclipses of the moons occur, and more. The other contains careful drawings of the Jovian system as seen through a telescope. This certainly does not look like a work full of falsehoods, written by an ignorant person who thinks things up to serve his own purposes while ignoring things as they are.

The combination of *Disquisitions'* many large and intriguing illustrations, Galileo's attention to it, and its relatively short length invites a reading—or, as the case may be, a translation. Modern readers who proceed beyond a casual perusal of *Disquisitions* will find that indeed it is not at all as Galileo portrays it, and not what one might expect from an anti-Copernican work. If what one expects from an anti-Copernican work is (to borrow some phrases from Albert Einstein's foreword to Galileo's

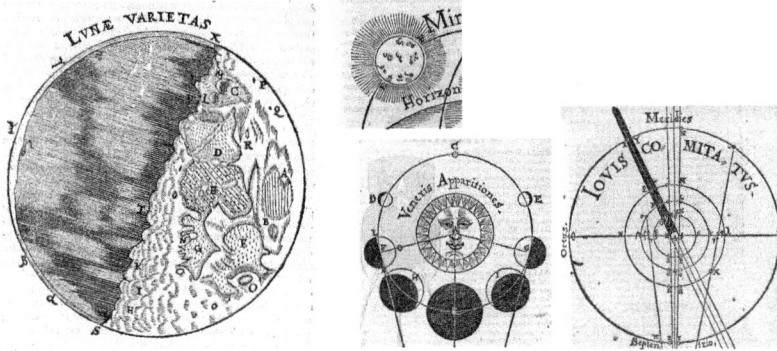

FIGURE I-1. Locher's illustrations of *(from left to right)* the moon, the sun (with spots), the phases of Venus showing that it circles the sun, and the Jovian system. All of Locher's illustrations used in this book are courtesy of ETH-Bibliothek Zürich, Alte und Seltene Drucke.

Dialogue) anthropocentric and mythical thinking, and opinions that have no basis but authority—against which Galileo stands as a representative of rational thinking[9]—then Locher's *Disquisitions* in fact invites a reevaluation of that expectation.

Locher seems adept at rational thinking. He begins with an excursion into mathematics, emphasizing how it "is ageless, unchanging, and certain. Nothing stands in opposition to it. It yields to no difficulties of philosophy. It deals in no tricks." He separates astrology—which he says is "speculation that seeks to divine or judge the influence of heavenly bodies on earthly events, and to gain insight into future events based on the positions of the stars and planets"—from astronomy. Astronomy, he says,

is more deliberate. It is that which studies absolute and inherent qualities of the heavens—number, shape, position, motion, time of occurrence, time of duration, qualities of light such as color or brilliance, and so forth.... It records and preserves celestial phenomena. It is the one friend with whom the heavens share their secrets. Elegant geometry and subtle arithmetic give it wings. Its paths become known to those who faithfully and carefully, through long and repeated experience, come to know its phenomena. Fine craftsmanship sustains their hands and strengthens their arms. Keen optics sharpen their eyes.

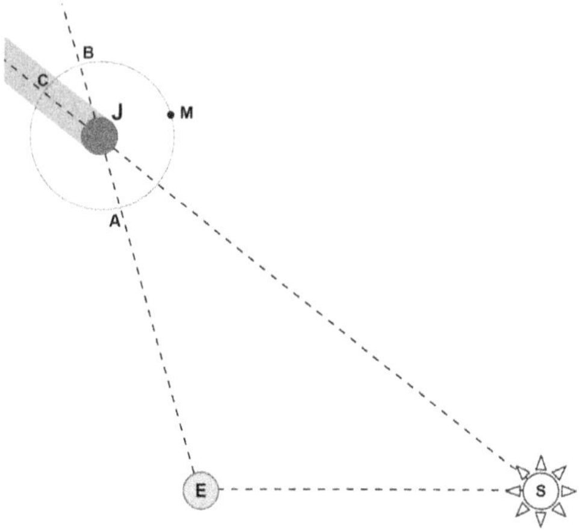

FIGURE I-2. Locher proposes using timings of the moons of Jupiter to measure distances between Jupiter, the sun, and Earth. Jupiter (J) casts a shadow that extends opposite the sun (S). A Jovian moon (M) circles Jupiter counterclockwise. An observer on Earth (E) notes by means of a telescope the time required for the moon to pass from the point at which it is in the center of Jupiter's shadow (C) to the point at which it is seen to stand directly in front of Jupiter (A). The ratio of that time to the period for one complete orbit of the moon is the same as the ratio of angle CJA to 360°. Thus angle CJA can be determined. Angles CJB and SJE can then be calculated using basic geometry. The angle SEJ between the sun and Jupiter can be directly measured from Earth. Since two angles, SJE and SEJ, are known in the sun-Earth-Jupiter triangle, and one side of that triangle (side ES) is one solar distance, the other two sides can be calculated in terms of solar distances, using basic trigonometry. Thus these distances can be directly determined, with no reliance on assumptions about the structure of the planetary system.

Thus Locher endorses what keen optics and skillfully constructed instruments reveal, and graphically and accurately represents that to his readers through the aforementioned illustrations.

Yet he goes further. Readers of *Disquisitions* will find that Locher proposes two research projects for the astronomical community. First he proposes that the newly discovered moons of Jupiter be used, together with geometry, to determine the distances between Jupiter, the sun, and Earth. Determining a certain angle in the Jovian system is key to this, he says (see Figure I-2), and "that in turn requires exact knowledge of the first emergences of the satellites from the shadow of Jupiter . . . after they have been eclipsed. This will require diligent and frequent observations." Then he proposes that the "attendants" of Saturn (not yet identified as rings) can be used to probe its orbital motion (Figure I-3). He says that "To find out what actually happens and settle these matters . . . Saturn must be diligently examined. . . . But we suspend judgment for now as regards all these matters of Saturn, and leave them to be decided by further experience with the phenomena."

Locher even advances a physical explanation for the phenomenon of Earth's motion around the sun in the Copernican system, namely that Earth is perpetually falling around the center of the universe, toward which it gravitates. In this way, he says, "we may be able now to imagine a manner by which Earth might be made to revolve" around the sun, even though he does not believe Copernicus to be correct. Indeed, students in introductory physics courses everywhere learn that Isaac Newton explained orbits as being a continual fall under the influence of a central gravitational force. The details are somewhat different, but Locher has the general idea.

Readers who delve into *Disquisitions* thus find Locher emphasizing the importance of mathematics, of long observation, and of recording data on position, motion, time of occurrence, time of duration, color, brilliance, and so forth. Readers find Locher creatively addressing interesting scientific questions, even about ideas with which he does not agree. Locher recognizes potentially productive research projects and encourages fellow astronomers to undertake prolonged efforts to gather the data needed to address these projects and answer certain questions, while holding off judgment until the results are in. In short, modern readers find Locher to be acting much like a modern astronomer, scientist, and rational thinker and not much like the exemplar of anti-Copernican silliness that Galileo portrays him to be.

Readers will also find that Locher displays a high regard for Galileo.[10] He is quite complimentary toward Galileo, a Copernican. At the same time Locher is extremely dismissive of Simon Marius, a fellow anti-Copernican. Locher's opinion seems to be that Galileo is outstanding, skilled, and learned, while Marius is, at best, a Galileo emulator.

At this point readers may wonder why Locher is not a Copernican, if he is a rational scientist, is friendly toward Galileo, and can even put forth

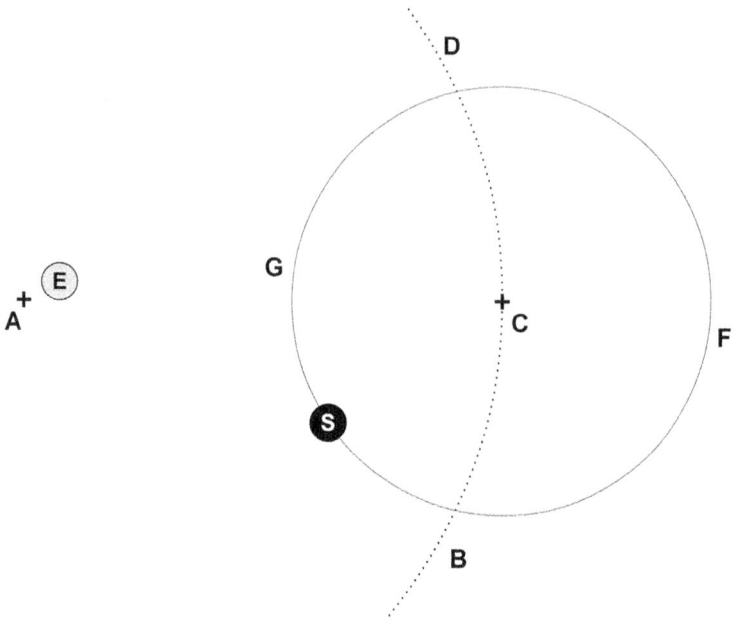

FIGURE I-3. Saturn is seen to slowly drift eastward through the constellations of the zodiac, but it periodically slows, stops, moves westward, and then stops again before resuming an eastward motion. While making the westward or *retrograde* motion, it grows brighter. The second-century Egyptian astronomer Claudius Ptolemy had explained retrograde motion *(above)* by supposing Saturn (S) to ride on a circle called an *epicycle*, which in turn rides on a larger circle called a *deferent*. The deferent is *eccentric* to Earth: its center (A) does not coincide exactly with Earth (E). Saturn is carried clockwise on the epicycle, going around once in roughly one year, while during that time the center of the epicycle moves on the deferent from B through C to D. The combination of motions means that Saturn generally moves clockwise relative to Earth, but when closest to Earth (at G), and therefore brightest, its motion relative to Earth is reversed.

a prescient explanation for how Earth could move around the sun. Why does he devote pages to arguments against the Copernican system, even when, as he puts it, so many astronomers of his time are burning incense at the altar of Copernicus?[11] Because, he says, "we follow reasoning and experience, and we are by no means easily swayed by assertions."

Modern readers of *Disquisitions* know that Copernicus was right and so may assume that reasoning and experience (observations, data collec-

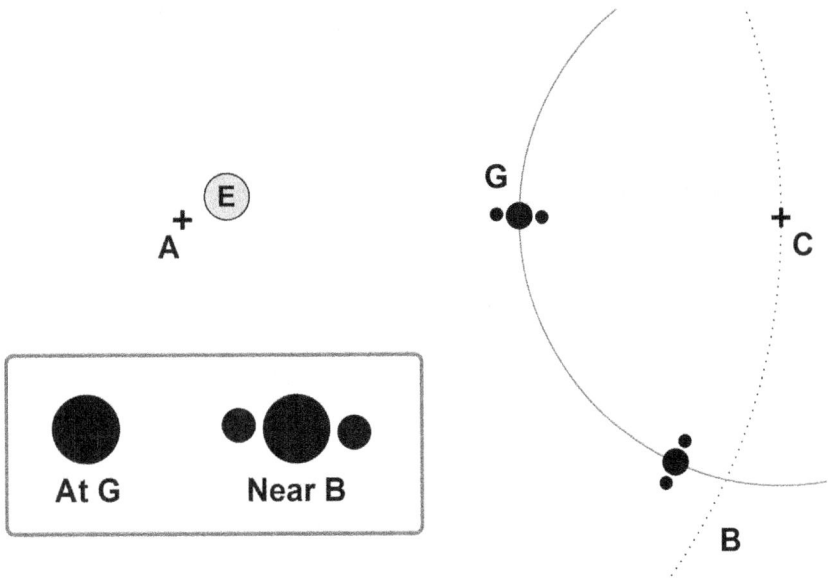

FIGURE I-3 (*cont.*). Galileo, using the telescope, had discovered Saturn to have two "companions" or "attendants," which had disappeared in 1612 and later reappeared. Locher argued that these attendants were probably not physically disappearing, but rather that this was an appearance caused by motion. He hypothesized that the companions might be locked to Saturn's epicycle *(below)*, so when Saturn is at G, the companions are in front of and behind the planet as seen from Earth and not apparent *(see insert, left)*, but when Saturn is near B they are visible *(insert, right)*. He proposed that astronomers engage in careful observation of Saturn and its companions over time to test this hypothesis, which at the same time would probe for the existence of a Saturnian epicycle. (Today we know that the "attendants" were Saturn's rings, poorly seen, and their temporary disappearance stemmed from them being edge-on to Earth in the summer of 1612.)

tion, calculations, etc.) would immediately lead to the right world system. The telescope proved the sun to have spots, Venus to circle the sun, and Jupiter to have moons that circle it. Clearly the telescope proved *wrong* the old Ptolemaic ideas about heavenly bodies being formed from a perfect celestial substance, about Earth being the center of all, and about epicycles and eccentrics explaining the motions of planets across the heavens (see Figure I-3). Readers' assumptions are encouraged by statements such as

Einstein's, or those found on the cover of a modern translation of Galileo's *Dialogue* describing Galileo "proving, for the first time, that the earth revolves around the sun."[12] According to this presumably reliable source, Galileo proved the matter. Considering that Locher illustrates in detail what the telescope reveals about the sun, Venus, and Jupiter, readers will certainly wonder how Locher does not see that proof.

Locher, readers will find, sees matters differently. To Locher, telescopic observations of the sun, Venus, and Jupiter have proved *right* the key Ptolemaic idea of epicycles: "the optic tube," he says, "has established that the center of Venus's own motion is the sun, that the center of the motions of the Jovian satellites is Jupiter, and that the center of motion of the solar spots is again the sun. Therefore epicycles do exist in the heavens." The second-century astronomer Ptolemy had *postulated* epicycles to exist because they explained the motions, as seen from Earth, of the planets (the "wandering stars"). Locher notes that the telescope now *proves* that epicycles exist, and that further telescopic study could reveal whether Saturn in particular is on an epicycle (see Figure I-3). Within the limits of the knowledge of his time,[13] Locher is correct. The motion of the Jovian moons is indeed epicyclic; the moons move on their orbital circles around Jupiter while Jupiter in turn moves on its own orbital circle.

Moreover, readers will find that Locher produces good reasons, following reasoning and experience and ignoring assertions, to reject Copernicus. He gives six arguments against Copernicus. The first is that the Copernican system inverts the words of astronomy (so that, for example, the sun doesn't rise, but rather Earth's horizon drops) and of Scripture (so that, for example, when Joshua told the sun to stand still, it was Earth that stood still). But Locher then retracts this first argument. The Copernicans can answer it, he says. Their answer might be convoluted, but it is satisfactory. Thus he gives five real arguments—ones to which the Copernicans have no satisfactory answer. All five are matters of science and reason. Not one relies on authority or mythical thinking.

Three of the five pertain to the physics of heavy falling bodies—to the question of how it can be that, on a rotating, spherical Earth, a heavy falling object is seen to drop vertically. The question is not a simple one. It almost overwhelms the pre-Newtonian, Aristotelian physics of Locher's time, but Locher is able to make his point: a rotating Earth transforms a simple fall into an incredibly complex phenomenon. Is not an immobile,

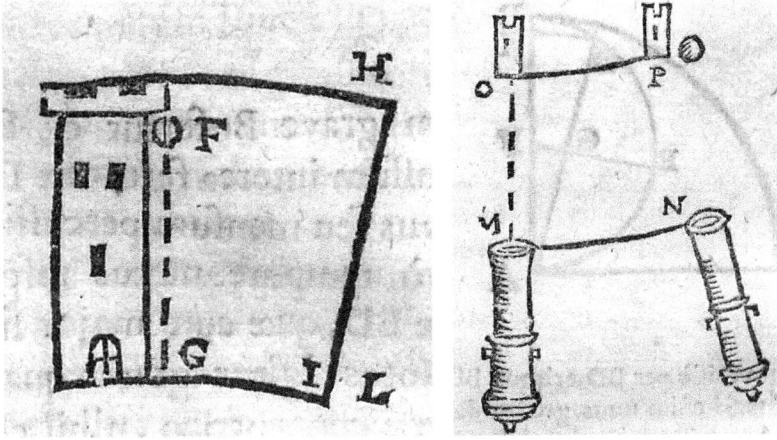

FIGURE I-4. The Coriolis effect. The top of a tower located near the Earth's equator is farther from Earth's center than the bottom of that tower. On a rotating Earth the top moves through a larger circle than, and thus faster than, the bottom.

Left: If Earth is stationary, then a ball dropped from the top F should fall straight down to the bottom G. If Earth is rotating, then as the ball falls the top moves to H while the bottom moves to I; but the ball, which is moving to the right at the speed of the top, should outrun the bottom and land at L. Thus, on a rotating Earth a falling ball should not drop straight down. Compare to Locher's disquisition 14. *Right:* Similarly, a projectile launched from the equator toward a target to the north should outrun the target and deflect to the right.

The Jesuit astronomer Giovanni Battista Riccioli developed this idea in the mid-seventeenth century as an argument against a rotating Earth (Graney 2015, 118–20; Graney 2011). These figures are from a later seventeenth-century Jesuit, Claude Francis Milliet Dechales; he used them likewise, as part of an argument against Earth's motion (Dechales 1690, 328). However, the effect does exist. It is the source of the rotation in hurricanes, among other things, and now bears the name of Gaspard-Gustave de Coriolis, who described it mathematically in 1835. Images courtesy of ETH-Bibliothek Zürich, Alte und Seltene Drucke.

central Earth, toward whose center heavy things naturally gravitate along straight lines, a far more simple and elegant solution to the question of falling bodies? Indeed, a description of a fall on a rotating world is not at all simple, even using modern Newtonian physics and the tools of differential equations; it involves terms such as *the Coriolis effect* (see Figure I-4).

Locher's remaining two arguments against Copernicus pertain to the stars, and in particular to the distances in the Copernican system of the

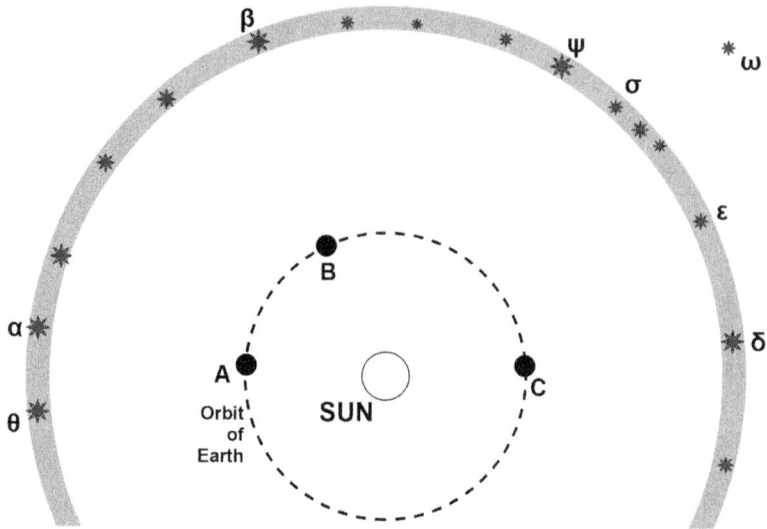

FIGURE I-5. Annual parallax. If Earth orbits the sun, moving from A to B to C within a six-month period, then a number of observable changes should be seen in the appearance of the stars. For example, when Earth is at A, star α will be closer and thus should appear larger than star β; two months later, when Earth is at B, the situation will be reversed. In short, the apparent size (magnitude) of a star should vary as the months pass. Or, if the stars extend into space, then stars σ and ω will appear close to one another in the sky when Earth is at B, but less so when Earth is at C. In fact, no such changes occur in the stars. The Copernican answer to why these sorts of changes are not seen is that the orbit of Earth is of negligible size compared to the distance to the stars: thus these annual parallax effects exist, but are negligibly small.

stars from Earth. These distances are vast compared to the distances of the sun, moon, and planets. If Earth circles the sun, then it moves relative to the stars and that movement should be reflected in the stars. But this effect, known as *annual parallax* (see Figure I-5), could not be detected in Locher's time. Indeed, it would not be detected until the nineteenth century. Copernicus attributed the lack of detectable annual parallax to the stars being so far away that the circle of Earth's orbit was vanishingly small by comparison, so any annual parallax would also be vanishingly small. By contrast, in a geocentric system the stars lie just beyond the planets, so the distances to all celestial bodies in such a system are comparable.

FIGURE I-6. An observer O on Earth sees a star as having a certain apparent size (indicated by arrows). The farther away the star is, the larger its true physical size must be in order to present that apparent size to the eye. If the star is located farther from Earth (at 2), then its physical size is much greater than if it is located closer to Earth (at 1). The heliocentric system required stars to be at vast distances from Earth and therefore to be enormous.

One of Locher's two star arguments is that the vast Copernican stellar distances serve no purpose. These distances are what we might call simply an ad hoc idea, introduced in order to make the heliocentric system conform to observations. Of course, we know today that Copernicus was right, but we can see Locher's point.

Locher's other star argument is based on the fact that stars have an apparent size.[14] Tycho Brahe determined, for example, that the more prominent stars appear about one-fifteenth the apparent diameter of the moon. Thus, were a prominent star of the same physical size as the moon, it must be about fifteen times farther away than the moon; were it of the same physical size as the sun (which has the same apparent diameter as the moon), it must be about fifteen times farther away than the sun. The farther away the star is, the larger it must be (Figure I-6). As Locher points out through various calculations, at the distances the Copernican system requires, that prominent star would have to be huge—far larger than even the sun. By contrast, in a geocentric system, where the stars are not so far away, the physical size of a prominent star would be comparable to the other celestial bodies. This argument was not Locher's; it was Tycho Brahe's primary anti-Copernican argument (Figure I-7).

Locher then reduces Brahe's star-size argument to a simple, elegant point: even the smallest star has some apparent size. It occupies some measureable fraction of the dome of the sky. Thus even the smallest star, tiny though it may be, is not vanishingly small compared to the sphere of the stars. By contrast, the Copernican theory requires Earth's orbit to be

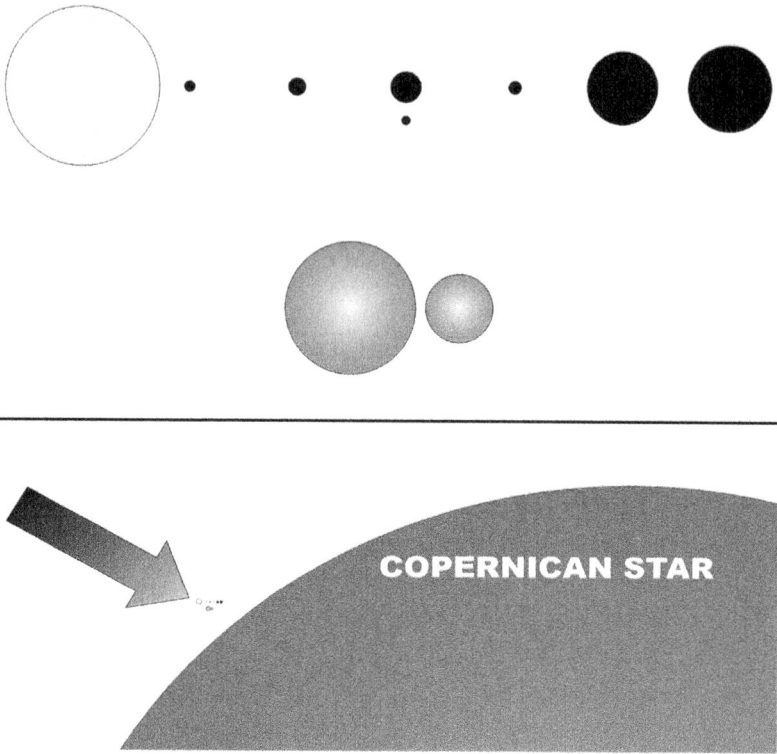

FIGURE I-7. Tycho Brahe calculated the physical sizes of stars required by both Copernicus's heliocentric system and Brahe's own geocentric system. A moderately prominent star appears to the unaided earthbound eye to be about the same size as Saturn. In a geocentric system, stars are just slightly beyond Saturn. Therefore, being about the same distance away as Saturn and about the same apparent size as Saturn, they turn out to have physical sizes comparable to Saturn. Shown in the upper figure are Brahe's calculated sizes for *(top row)* the sun, Mercury, Venus, Earth and the moon, Mars, Jupiter, and Saturn, and *(bottom row)* a prominent star and a modest star. Thus, in a geocentric system, all celestial bodies fall into a reasonable range of sizes, with the sun being the largest and the moon the smallest.

In the heliocentric system, by contrast, stars lie hundreds of times farther away than Saturn and therefore must be hundreds of times the diameter of Saturn. Shown in the lower figure is a modest star in the Copernican system, with the upper figure at its left for comparison. Diagrams from Graney 2013.

vanishingly small compared to the sphere of stars in order to explain the absence of parallax. *Small but measurable* is larger than *vanishingly small,* and therefore every last star must be larger in size than the orbit of Earth. Thus every last star must utterly dwarf the sun. Thus the Copernican theory does not just require the stars to be at vast distances, it requires them—every last one—to be spectacularly huge. The Copernicans, says Locher, do not deny this, but say all this testifies to the power of God. (And indeed, one prominent Copernican would eventually go so far as to declare the giant stars to be God's mighty warriors, the palace guard of heaven itself, and to support this notion with abundant quotations from Scripture.[15])

This is why Locher sees reasoning and experience as being contrary to Copernicus. This is his reasoning for not accepting the Copernican system, despite his detailed illustrations of what the telescope reveals, and despite even his explanation for how Earth could move in that system.

Modern readers of *Disquisitions* will find that Locher covers a number of other topics, also investigated in a thorough and rational manner wholly at odds with Galileo's portrayal of the "booklet of theses." These include the appearances of the moon and of small light sources seen at a distance, and the effect of the atmosphere on the appearance of heavenly bodies. Locher even discusses a topic that seems amazingly modern: the question of whether our universe is just one of an infinite number of universes randomly forming from and dissolving back into elementary particles within an infinite universe of universes (that is, within a multiverse, to use a modern term). He argues against the multiverse idea by means of a variety of mathematical arguments disputing the existence of any physical infinity. Today arguments against the existence of a multiverse, or of a physically or temporally infinite universe, have more than a little in common with those of Locher—usually arguing that infinities require all possible events, no matter how improbable, to occur within the infinity, and to occur infinitely often, and that postulating infinite numbers of improbable events (for example, infinitely many other universes in which this book is set with pink type, or purple type, or rainbow-colored type) to explain the existence of our universe represents an infinite violation of Ockham's razor. And, much like those who engage in such arguments today, Locher sees in the disproof of the multiverse evidence for the existence of God.

Thus readers of Locher's *Disquisitions* will find that it contains much that is interesting. It defies both Galileo's portrayal of it specifically and expectations of anti-Copernican works generally. It also defies modern portrayals. Even the best modern scholars have described *Disquisitions* in ways that are at variance with the contents of the book itself—describing it as violently attacking Copernicans, as opposing Galileo's descriptions of the moon as seen telescopically, or as putting forth trivial objections to heliocentrism.[16] No doubt this variance, which is always on the negative side with respect to Locher's work, is owed to the fact that Galileo himself conveyed a negative impression of this "booklet of theses."

Anti-Copernicans in general may be portrayed too negatively.[17] They backed a hypothesis that turned out to be wrong, and they faded into obscurity. Galileo backed a hypothesis that turned out to be right, and he became one of the most famous figures in the history of science. Galileo was not one to portray his opponents in a positive light, but of course in Galileo's time, his opponents had not yet faded into obscurity and could speak for themselves. In part because Galileo chose Locher as an example, immortalizing him in the *Dialogue*, modern readers can delve into *Disquisitions* and let Locher speak for himself. They can get a better sense of the scientific dialogue in Galileo's time—a dialogue that may seem, after reading *Disquisitions*, less like Einstein's rational thinking standing against authority and anthropocentric and mythical silliness, and more like a true scientific dialogue.

NOTES

1. Galilei 2001, 105.
2. Galilei 2001, 107.
3. Galilei 2001, 108.
4. Galilei 2001, 414.
5. For example, see Van Helden and Reeves 2010, a book specifically on Scheiner that speaks of Locher's work as though it were Scheiner's (307–8); see also Drake 1958, 157–58, and Heilbron 2010, 275–76, 436. From the beginning some have attributed *Disquisitions* to Scheiner. Galileo does not mention *Disquisitions* or its author by name, but Giovanni Battista Riccioli, in his *Almagestum Novum*, refers to the book and cites Scheiner as the author (Riccioli 1651, 54). An

interesting published reference to the *Disquisitions* that describes it as being Scheiner's is a 1793 work by Giovan Battista Clemente Nelli that mentions how "Jesuit Father Christopher Scheiner . . . composed a miserable brochure against the motion of the earth" (Nelli 1793, 417), namely *Disquisitions*. Nelli cites a letter of August 27, 1616, from Giovanni Francesco Sagredo to Galileo. Sagredo's letter was published by 1856 (see Galilei 1856), and after this time references to *Disquisitions* are more common and generally attribute the work to Scheiner. Sagredo writes, "Twenty-three months ago a book was sent to Sr. Magini, printed in Ingolstadt, entitled *Disquitiones mathematicae de controversiis et novitatibus astronomicis*," which he says he borrowed out of curiosity since he understood it to attack the Copernican system. He goes on to say that Galileo should look at it if he has not done so, it being "the work of Jesuit P. Cristofforo Scheiner, who is that friend of Sr. Velser, whose head you once washed without soap, because of the disrespectful manner in which he wrote of me. . . . I've only read the smallest part of it, having other business now, nor do I find satisfactory the teachings of this most pretendentious man" (Galilei 1856, 112). (I thank Lorenzo Smerillo and Roger Ceragioli for their assistance with translating Sagredo's letter from Italian via the HASTRO-L history of astronomy listserv on May 4, 2016. According to Smerillo, the "head wash" idiom is still in use in Italian.)

6. Galilei 2001, 60–63, 402–13, 548–49. Also see Van Helden and Reeves 2010, 321–27.

7. Finocchiaro 1989, 37, 265.

8. Galilei 2001, 80.

9. Galilei 2001, xxiii, xxviii.

10. Drake 1958, 158.

11. Locher notes that fewer people ascribe to a geocentric system in his time than in ancient times, "since the majority [in his time] burn incense to Copernicus" (cum plerique Copernico turiant). The Latin *tus* is "incense" and *turifico* is "burn or offer incense." See Locher 1614, 23.

12. Galilei 2001, back cover.

13. We all are constrained by the limits of the knowledge of our time, as students living in the twenty-fifth century will undoubtedly learn when they study early twenty-first-century astronomy.

14. Today we know that these apparent sizes are an illusion caused by the nature of light and that stars seen from Earth are actually point-like sources of light, entirely consistent with their being sun-like bodies at Copernican distances. How astronomers came to discover this fact, and therefore the solution to the star-size argument, is discussed at length in Graney and Grayson 2011 and in Graney 2015, 148–57.

15. Graney 2015, 63–85; Graney 2013.

16. Various writers have characterized *Disquisitions* in ways that are at odds with the contents of the book, and surely one reason for this is the influence of Galileo's characterization of it. See Finocchiaro 2010, 71; Finocchiaro 2013, 132; Reeves 1997, 205, 206, especially vs. Van Helden and Reeves 2010, 309; Heilbron 2010, 276; and Piccolino and Wade 2014, 132.

17. Giovanni Battista Riccioli has regularly received an inaccurate negative portrayal in secondary sources: see Graney 2015, 103–5.

Translator's Note

This translation of Locher's *Disquisitions* renders this Latin mathematical and astronomical work into English for students in even introductory classes in history, physics and astronomy, or history and philosophy of science. As such, I intend it to be an accurate reflection of his work that is readable by a broad audience, especially students (including advanced high school students) who might be assigned reading from Galileo's *Dialogue* as part of these classes. I thus err on the side of readability, often rendering technical astronomical or metaphysical terms in a manner that explains the terms. In the technical mathematical sections I at times significantly reorder or shorten Locher's original phrasing into a structure that is more familiar to modern readers, while taking care to ensure that Locher's mathematics itself is unaltered. Locher uses many marginal notes, usually to cite sources, but also for emphasis. Typesetting these as marginal notes is impractical, so I incorporate their content into the main text and indicate it with ‹pointed brackets›. Locher's figures are unlabeled and simply appear within the appropriate text. As a practical matter I have numbered the figures (e.g., the first figure in his fourth disquisition would be Figure 4-1, the third figure in his eighth disquisition would be Figure 8-3) and insert text into the translation directing readers to the figures by number. Inserted text I indicate with [square brackets].

This translation is not intended for expert scholars, although they should find it useful. Scholars interested in Locher's precise word use, marginal note structure, figure placement, and so on can easily consult the original *Disquisitiones mathematicae*, freely available online via, for example, E-rara and Google Books. (Indeed, thanks to the nature of technology, Locher's original is today more readily available to scholars than is this translation.) However, this translation should be an excellent resource for a scholar wishing to produce a scholarly rendition of Locher's original words.

There are two areas in which my approaches to the translation differ from what I have just described. One is the poems that come before and after Locher's work. These I have rendered extremely loosely—these renditions might better be described as words inspired by the original Latin. The other is those places where Locher directly quotes other authors. There I have used existing translations when available and attempted to produce close translations (with less concern for readability) when existing translations were unavailable. At times this means that such quotations contain awkward language, but the reason for doing this is to prevent all the material (Locher's words, and the words of the authors whom Locher quotes) from sounding the same. If I translated everything in the same manner, then both Locher's words and the material he quotes would have the same "voice"—that of their translator.

The Structure of the *Disquisitions*

This is an English rendition of the entirety of Johann Georg Locher's *Mathematical Disquisitions Concerning Astronomical Controversies and Novelties*; the only material not rendered into English is the list of typographical errors that occupies part of the last page. Therefore readers will encounter within this book more than just the forty-four disquisitions that comprise the heart of the book. There are three laudatory poems, one at the beginning and two at the end. There is a letter of dedication from Locher to his lord and another letter on the value of mathematics.

Then there are the disquisitions themselves. The early ones address the nature of mathematics and are generally brief. Astronomy begins to be discussed by disquisition 7. By disquisition 13 Locher is addressing the Copernican system and arguments against it, and the disquisitions grow much longer. He moves on to discussing other systems and the problems with them. By disquisition 25 he has begun discussing the specifics of different celestial bodies, starting with the moon. It, the sun, Venus, and Jupiter and its moons (with related excursions into subjects such as light and the effect of the atmosphere on observations) comprise the rest of the disquisitions, except for the last. That short disquisition, number 44, concerns Saturn and its rings (not yet recognized for what they are).

DISQVISITIONES
MATHEMATICÆ,
DE CONTROVERSIIS ET NOVITA-
TIBVS ASTRONOMICIS.

Quas

SVB PRÆSIDIO
CHRISTOPHORI
SCHEINER, DE SO-
CIETATE IESV, SACRÆ LIN-
GVÆ ET MATHESEOS, IN ALMA
Ingolstadiensi Vniuersitate, Professo-
ris Ordinarij,

PVBLICE DISPVTANDAS
POSVIT, PROPVGNAVIT,

Mense Septembri, Die

NOBILIS ET DOCTIS-
SIMVS IVVENIS, IOANNES
GEORGIVS LOCHER, BOIVS MO-
NACENSIS, ARTIVM ET PHILO-
sophiæ Baccalaureus, Magisterij
Candidatus, Iuris Stu-
diosus.

INGOLSTADII,

Ex Typographeo Ederiano apud Elisa-
betham Angermariam.

ANNO M. DC. XIV.

* * * * *

MATHEMATICAL DISQUISITIONS, CONCERNING ASTRONOMICAL CONTROVERSIES AND NOVELTIES

The noble and learned young man,

JOHANN GEORG LOCHER

(Bavarian of Munich, bachelor of arts and of philosophy,
master's candidate, student of law)

has publicly put forth for discussion and defended these,
under the supervision of Christoph Scheiner,
of the Society of Jesus,
ordinary professor of sacred language and mathematics,
at the University of Ingolstadt.

From the Ivy Press of Elizabeth Angermaria.
Ingolstadt, 1614.

To the most serene prince and lord,
Lord Maximilian, imperial count of the Rhine,
duke and most clement lord of Bavaria, etc.

The new, the rare, and the precious are owed to the prince, O mighty Duke. Custom has sanctioned this. Reason has advised this. The consensus of everyone has confirmed this. This is true not only in your Bavaria (where items that are of extraordinary value, or that are novelties, or rarities, or works of great craft have been gathered into and guarded within your own archive of special wonders in Munich), but indeed in all civilized nations.

Old coins dug out of the ruins at Rome soon reach the hands of the mighty. Monsters caught from the sea — the kind that once inspired poets to write of the Sirens — are promptly sent to the palace. Kingdoms in India fight among themselves for the white elephant, not because it is better than the dark elephant, but because it is rarer. Clovis II, the greatest king of the Gauls, even acquired a turnip of unusual size.

And now I entrust to you a rarity that is even lowlier than a turnip. I have published these mathematical theses concerning celestial novelties. Because these theses consider celestial things, perhaps they are not too lowly — after all, it is probably better to calculate about heavenly things based on philosophical opinion than to be learned about other things.

3

Because these theses consider things of fascinating rarity and abounding in novelty, perhaps they have worth.

I owe much to the professors of mathematics at your academy at Ingolstadt, some of whom I have acknowledged publicly, others of whom I have thanked privately. To both, I give this work as a sign of a grateful heart. And if you do not spurn it, then perhaps in the future this rarity may not seem so rare.

Live long, and prosper, God willing—to you, to our country, and to us. Ingolstadt 1614.

All the best,

your most dutiful servant in all things.

Johann Georg Locher

A Poem for the Most Learned Reader

Orpheus, put the Argonauts out of your mind. Put aside your lyre that once moved rocks and stilled waters. Hang the Golden Fleece back on the tree. Forget witches and the voices of Sirens. You may again rescue your beloved Euridice from Hades—for now you can pull the moon down from heaven and then watch Hades open wide in astonishment.[1] The world of the ancient gods is no more, except for in poetry and Scripture. The old ground on which the heavens have shone now trembles.

Now Cynthia the Moon shines only by the sun's fire, yet still she leads the stars through the night. The starry multitude celebrates God; the Milky Way is full of them. Jupiter drives a chain of four golden stars, while Saturn can but envy Jove. Shining Venus has no pristine form to her face. Instead, her brow bears two brazen horns. How fitting that she, whom the fool deems most pretty, in the end prefers the foolishness of these tasteless horns.

And the bright-shining phaethonic lamp—Phoebus the Sun—also bears a marred face. His face is branded with dark burns, an astral stigma that brings to mind the crime of Prometheus, who gave fire to man. But lest Phoebus be merely an orb darkened by spots, he is also whitened by numerous blazes.

5

When I consider all this, I leap up and cry, O heavens, what craft is this, when I find day and night, bright and dark, on the sun itself? What is this labor that drives the one who would seek dark spots to study solar fire?

Yet go on, look up at the stars, and summon them down. Argus,[2] why are you not indeed a lynx? I say you are above the lynxes. Others may be lynxes. To me you are an eagle.

Henricus Locher, brother to brother,
student of logic and mathematics.

The Author to the Reader

On the Preeminence, Necessity, and Utility of Mathematics

Dear Reader, lest I seem not credible, or too eager to press my case, I shall give to you not my words and opinions, but those of others. Thus I give you the words (abridged for length) of that great theologian Antonio Possevino,[3] from his Bibliothecae Selectae. *He is a better witness for such things in all ways. ‹In the "Argument" section of book 15›, he writes:*

> *I am persuaded, and I have proposed to others, that the mathematical disciplines have a place and a dignity ahead of the majority of the higher disciplines. This is because the mathematical disciplines are necessary not only to the rest of the sciences, and to medicine, but also very much to the conduct of war, to various operations of the Republic, and to studies of geography and human history. . . .[4]*

Then we find the following ‹in chapter 1 of book 15›, which I have abridged somewhat for length:[5]

> *Those demonstrate the necessity, dignity, and utility of the mathematical disciplines. Plato and Aristotle have adopted these disciplines for a method*

of contemplating and doing. And certainly the best evidence of this is the Timaeus *of Plato and the* Physics *of Aristotle. They bring forward the light of philosophy by means of mathematics itself, etc.*

Truly Aristotle's whole account about motion and rest, about time and the heavens, and also about the development of animals—and indeed his entire physical discussion—abounds not so much in examples as in geometrical foundations. Indeed, by the middle of the first book of Physics *Aristotle brings up Antiphon's "squaring of the circle," so that he may reject it.[6] In the second book he discusses the two right angles in a triangle,[7] which he also discusses in his* Posterior Analytics.[8] *In the third book he mentions some things about building gnomons, and then the remainder concerns infinity of size, motion, and time. Most people introduced to the ideas of Aristotle lack solid understanding of these books because they have never deeply perceived the mathematical disciplines.*

And in the book On the Heavens, *because a diameter is not comparable to a side, Aristotle discussed a sphere constructed around eight pyramids, and thus he discussed pyramids, and thus triangles.[9] His book* Meteorology *is full of mathematics.[10] The same must be said regarding* Metaphysics. *Indeed, book 12 of* Metaphysics *considers whether mathematics is a real thing, whether numbers are real things, and whether mathematical ideas are the most fundamental of all. Thus a certain knowledge of mathematics is a necessity.*

Mathematics also pertains to theology. For example, settling the date of the celebration of Easter—and of the rest of the "moveable feasts," as they are called—was a concern both at the ancient synod at Nicaea[11] and at the most recent synod at Trent.[12] The order and management of the whole Christian Republic is arranged according to that date. I shall not digress to those things which occur everywhere in the Scriptures regarding the stars and the heavens, the measurements and architecture of the temple of Solomon,[13] and countless other things.

If we consider medicine, we find Galen[14] stating that a doctor who is ignorant of the timing and duration of proper treatments must not treat the sick. By such ignorance a doctor may bring a patient to ruin, rather than to the health and soundness the patient might expect. Likewise, a farmer who ignores the proper timing of grafting, transplanting, and sowing will usually experience want and require charity.

Plutarch[15] reports in his Life of Marcellus *how Archytas and Eudoxus[16] added variety to geometry by removing it from the realm of pure mental exercise and bringing it into the realm of real and practical things,[17] which are now found in Aristotle's* Mechanics.

The fruits of such practical applications of mathematics are many. Archytas created a flying wooden pigeon.[18] Archimedes and Posidonius[19] constructed working mechanical models, or planetaria, that replicated the movements of the sun, moon, and planets—this is reported by Cicero,[20] who notes that in making these planetaria they replicated the action of God in building the universe,[21] as Plato describes in the Timaeus.[22] *In more recent times the Nürnberg artist, Albrecht Dürer, illustrated his fly and his eagle with geometrical wings. Claudius Gallus constructed for the gardens of Cardinal d'Este elaborate mechanical birds, driven by hydraulic action. Small copper birds would sing and move until a little mechanical owl appeared, and then when it departed they would resume their activities.[23] They were so realistic that a person who declared them fakes would seem more temerarious than a person who claimed them real birds would seem credulous.*

Other practical fruits of mathematics relate to measurement. A single measuring rod, used to measure distances, can also be used to determine the areas of surfaces and the volumes of bodies. Using a simple measuring rod, any geometer can describe buildings, lands, seas, the movements of the heavens, the risings of stars, and so forth.

But these things are not all that is encompassed by the discipline of mathematics. Plato wrote:

> *In dealing with encampments and the occupation of strong places and the bringing of troops into column and line and all the other formations of an army in actual battle and on the march, an officer who had studied geometry would be a very different person from what he would be if he had not.[24]*

Indeed, there are many military applications of geometry. In the Roman army, a centurion's flag served as a point around which a circular or rectangular formation of troops would be established, depending upon the circumstances of battle. Geometry guides the construction of bridges and ships, the channeling of water, and the movement of cavalry among foot soldiers. It can be used in both attack and defense—in the construction of both siege engines and defensive ramparts.

Geometry has influenced the outcome of a remarkable variety of battles. A small group of Caesar's soldiers broke through the ranks of a vast army by means of a wedge formation and escaped unharmed. Likewise, three hundred legionaries held back another vast army for hours through the use of a circular formation. At Syracuse, Archimedes constructed such defenses against Marcellus[25] that Marcellus called him "this Briarian[26] engineer and geometrician [who] hath with shame overthrown our navy, and exceeded all the fabulous hundred hands of the giants, discharging at one instant so many shot among us."[27] Zonaras[28] reports that Proclus,[29] by means of mirrors fashioned to collect and concentrate light from the sun, "burnt the fleet of Vitellius, at the siege of Constantinople, in imitation of Archimedes, who set fire to the Roman fleet at the siege of Syracuse."[30] And Archimedes and Proclus are but two of the people who have used geometry for military purposes.

Possevino was a noted theologian, but he was also a noted expert at law. Therefore, Reader, you will want to know that in a letter to a certain friend he writes,

It is right to acknowledge the importance to jurisprudence of the mathematical disciplines, especially arithmetic and geometry. Without arithmetic, who could apply Roman law in cases regarding inheritances and children born after the death of a father and not mentioned in his will? Such examples occur everywhere in law. And certainly no skill is more useful and more necessary to jurisprudence than knowledge of geometry. Who could settle matters of land ownership and titles without such knowledge? Who could make judgments in cases where sedimentation adds to land or causes new islands to arise, or where the changing courses of rivers alter the land? Likewise, how would a judge who lacks any skill in geometry ever determine whether a surveyor's measurements of a property are correct or incorrect?

Imagine a field of some sort being sold. There could be some contention that exists regarding the boundaries, or the buyer or seller may simply wish to know the boundaries. Perhaps the width of some right-of-way needs to be known. The labor required to work the land, the timber or other resources that can be extracted from the land, the grain or wine the land can produce, and the law as it pertains to these things—without cognition of geometry,

*judgment cannot be made about any of this. Furthermore, without geome-
try, no one can determine the rightness or wrongness of any judgment that
might be made. Indeed, geometry is no less necessary to the practice of the
law than is that basic logic which says, for example, that two opposing statements cannot both be true simultaneously.*

*For this reason it is well known that less prudent fathers push their sons
to the study of law before they are imbued by liberal education to some extent. Such men, eager for profit, save a year or two in the short run, but in
the long run they waste many years.*

*To these words of that most learned man I add this: Indeed, those who
are erudite at law, but truly ignorant of astronomical things, will confuse astronomers with astrologers—the prudent with the temerarious. They will be
unable to judge the testimony of a mathematician versus that of a crank.
They will condemn the innocent with the guilty, contrary to human and divine law. May they all come into a knowledge of astronomical things.*

*I might add a third statement of testimony (one to the usefulness of as-
tronomy specifically) from Christopher Clavius[31] in the preface of his treatise
on the astronomy of Johannes de Sacro Bosco.[32] However, he is so widely read
that there is no purpose to repeating it here. And Proclus discusses in chapter
8, book 1 of his book on Euclid how the same mathematics may be conducive
to political and moral philosophy, dialectics, rhetoric, poetry, and many other
things which are matters of art or are done through the hands and work.[33]*

*You have, Reader, the thoughts of my mind, through the words of oth-
ers—through their thoughts in their words. Had I expressed my mind
through my own words, they might have held less weight with you and perhaps have seemed less trustworthy to you. Tell me if you understand this
study I have made. If your understanding is greater than mine, go before me,
with Possevino—I shall eagerly follow you. If it equals mine, go with me—I
shall not refuse you. If it is lesser, follow me—I shall not impede you should
you overtake me. You have heard what I think. You see the point I wish to
make. You understand what I would like you to do.*

Farewell.

From my study at Ingolstadt. 15 August 1614.

DISQUISITION 1.

Quantity is the object of mathematical science. No one denies this. But no mathematician claims this to be complete or adequate.

Proclus states ‹in book 1, chapter 12 of his commentary on Euclid›,

> The Pythagoreans, therefore, thought that the whole mathematical science should receive a fourfold distribution, attributing one of its parts to the *how-many*, but the other to the *how-much*; and they assigned to each of these parts a twofold division. For they said, that discrete quantity, or the *how-many*, either subsists by itself, or must be considered with relation to some other; but that continued quantity, or the *how-much*, is either stable or in motion. Hence they affirmed, that arithmetic contemplates that discrete quantity which subsists by itself, but music that which is related to another; and that geometry considers continued quantity so far as it is immoveable; but spherics [astronomy] contemplates continued quantity as moving from itself. . . .[34]

And so, according to this, the object of mathematics is supposed to be not only quantity but also number, in either an abstract or a concrete sense, and so forth. Then ‹in chapter 13› Proclus states,[35]

> Again, some think (among whom is Geminus) that the mathematical science is to be divided in a different manner from the preceding. And they consider that one of its parts is conversant with intelligibles only, but the other with sensibles, upon which it borders; denominating as intelligibles whatever inspections the soul rouses into energy by herself, when separating herself from material forms. And of that which is conversant with intelligibles they establish two, by far the first and most principal parts, arithmetic and geometry: but of that which unfolds its office and employment in sensibles, they appoint six parts, mechanics, astrology, optics, geodæsia, canonics, and logistics, or the art of reckoning.[36]

Then in ‹chapter 3 of› the same work, Proclus writes that mathematicians must think about—

all considerations respecting proportions, compositions, divisions, conversions, and alternate changes: also the speculation of every kind of reasons ... together with the common and universal considerations respecting equal and unequal, not as conversant in figures, or numbers, or motions, but so far as each of these possesses a common nature essentially, and affords a more simple knowledge of itself. But beauty and order are also common to all the mathematical disciplines, together with a passage from things more known, to such as are sought for.[37]

From all these words it is sufficiently evident that the object of mathematics is not only quantity, which is limited to line, surface, and volume. Rather the object is much broader.

DISQUISITION 2.

Certainly, all that can be said in many ways. In one sense the object of mathematics, quantity, is the extension of a whole, of an entity, and not really distinct from the whole. Mathematicians have little concern for this. In another sense, quantity extends per se to be almost a thing itself, pertaining to a class of things.[38]

And although quantity may be a characteristic of a thing, nevertheless, of those characteristics a thing can have, it is noble to the highest degree. It approaches being a thing that exists of itself. It can be considered and used, like a thing that exists. It falls under no other subject. It is ageless, unchanging, and certain. Nothing stands in opposition to it. It yields to no difficulties of philosophy. It deals in no tricks. It is a certain succinct exhibition of divine power and immense wisdom. It is always finite actually, yet truly infinite potentially. Whether you increase by adding, or you diminish by dividing, you reach a certain numerical value. However, through adding or dividing an infinite multitude of values potentially can be reached. And mathematicians have admired this greatly and have honored this, and the ancients arranged everything on account of this.

DISQUISITION 3.

Quantity can be considered in three ways. Either there can be a whole that consists of discrete parts where each part can stand on its own, or there can be a whole that can be continuously divided into parts, and this in turn either according to a path or succession [like the division of time] or not [like the division of space].[39]

Mathematicians give over the first way [of discrete parts, of arithmetic] to the theologians and physicists, who merely touch on the parts of things and frequently make suppositions—like axiom 8, book 1 of Euclid.[40] The mathematicians generally weigh quantity in the other (continuous) way, concerning the measurement of or the ability to measure extension, or shapes, or the growth or division of this or that body, potential, quality, or virtue. All these things barely harmonize with that radical and in a certain way potential quantity,[41] strange to the ancient mathematicians, so the teachers of this discipline have very little to do with that.

DISQUISITION 4.

In metaphysics one abstracts from particulars to generals [in pursuit of the supreme feat of unifying all classes of reality together in relation to each other, and thus perfecting knowledge].[42] In physics the abstraction is from particular relations to general relations in material things—between the material that comprises things and the structures that govern that material, and within the structures themselves (considering structure clearly). In mathematics the abstraction is separated from the material things and considers the quantitative relations found in things. The metaphysicist considers being as a thing itself; the physicist considers relations in material things; the mathematician, in considering immaterial relationships, clearly walks the middle path between the other two, being closer to the material than the metaphysicist, yet further than the physicist. Thus Proclus ‹in chapter 10 of book 1 of his work on Euclid› writes,

> After the same manner, the mathematical science is indeed the second from the first of all sciences, and, with reference to it, imperfect: but it

is, nevertheless, a science, not as being free from supposition, but as knowing the peculiar reasons resident in the soul, and as bringing the causes of conclusions, and containing the reason of such things as are subject to its knowledge. And thus much for the opinion of Plato respecting mathematics.[43]

A little above this he writes,

We must not say, therefore, that Plato expels mathematical knowledge from the number of the sciences, but that he asserts it to be the second from that one science, which possesses the supreme feat of all: nor must we affirm, that he accuses it as ignorant of its own principles, but that receiving these from the master science dialectic, and possessing them without any demonstration, it demonstrates from these its consequent propositions.[44]

Therefore metaphysics has being itself as its object; mathematics has as its object what is possible; physics, what actually exists.

DISQUISITION 5.

Mathematics demonstrates her conclusions scientifically, through axioms, definitions, postulates, and suppositions. From this it is clear that mathematics should truly be called a science. If anyone would demand that a science treat physical matters, he must also exclude metaphysics from the ranks of the sciences, and even more so logic—and yet these are not excluded from the sciences.

DISQUISITION 6.

That stuff of mathematics that deals with establishing the higher and more universal principles, such as what treats the mathematical elements, especially emulates metaphysical abstraction. Euclid, in his books on the elements ‹(see Proclus book 2, chapter 2)›, thoroughly describes lines,

angles, shapes, sizes, and their limits, divisions, ratios, mathematical relations, equalities, applications, inequalities of greater and less, and the nature of numbers. He is not concerned with a triangle made from a physical element, such as earth or air or fire or aether, but with the triangle in
general, in the abstract sense, however that may be—not some sort of
utopian triangle, as some people incorrectly think, but the triangle as a
quantity that has existed since the beginning of time.

DISQUISITION 7.

Other mathematical sciences are less universal and deal (indirectly, not
directly) with particular objects and material things; with measuring their
size, shape, number, and so forth. Thus the accountant applies numbers
to money; the musician applies them to sound; the architect applies them
to symmetry, the configurations of buildings, and so on; and the astronomer applies the ratios of circles and spheres to celestial objects. So the
astronomer's principle objective, then, is quantity, as it pertains to the
attributes of a celestial body or to learning about a celestial body.

DISQUISITION 8.

It is commonly supposed that any sort of speculation concerning things
in the heavens is astronomy. However, speculation that seeks to divine or
judge the influence of heavenly bodies on earthly events, and to gain insight into future events based on the positions of the stars and planets, is
astrology.

Accepted astronomy is more deliberate. It is that which studies absolute and inherent qualities of the heavens—number, shape, position,
motion, time of occurrence, time of duration, qualities of light such as
color or brilliance, and so forth. It is the most noble and most ancient science, as old as humankind itself, almost an offshoot of the human soul
‹(see Proclus book 1)›,[45] and always held in the greatest esteem by all. It is
cultivated by kings,[46] the foundation of chronology, the basis of sundials

and of the computation of the liturgical calendar. It records and preserves celestial phenomena. It is the one friend with whom the heavens share their secrets. Elegant geometry and subtle arithmetic give it wings.

Its paths become known to those who faithfully and carefully, through long and repeated experience, come to know its phenomena. Fine craftsmanship sustains their hands and strengthens their arms. Keen optics sharpen their eyes. Labor supports them. Truth watches over them. Honor attends them.

And now, we shall display but a few pearls from the treasures of astronomy, this queen of the sciences. Let us begin with the question of whether astronomy might study the infinite.

DISQUISITION 9.

Every material thing is contained within the circumference of the heavens. Earth is perceived to stand at the center, greatly separated from the heavens. Around earth lies water, and above water, air, and above air, aether. The heavens envelop all these like a womb. The bulk of the universe consists of these things, from which all other things come together.

Many ancient thinkers ‹(see Plutarch's discussion on the universe)›, and many more recent ones as well, have opposed this arrangement of the universe and positioning of Earth. Many of the ancient philosophers and mathematicians believed in an infinite multiverse,[47] within which lies our universe and others. Each universe is finite in size, yet they are infinite in number. These lie within a great infinite chaotic swarm of elementary particles[48] that feeds existing universes and continually spawns new universes.

Today the structure of our universe is disputed, and so in [Figure 9-1] the part that would show Earth, sun, moon, and planets is left blank. Only the firmament of stars that encloses these is shown (A B C D). Beyond the firmament is the infinite chaos of particles. Our universe supposedly formed from this chaos, and swims in it now. And, supposedly, in time our universe will dissolve back into the chaos.

However, truth and Christian philosophy exploded these fictions some time ago, as we shall now demonstrate.

FIGURE 9-1

DISQUISITION 10.

First, it is actually impossible to have any infinity, either in multitude, or in magnitude, whether enclosed by a boundary or not. Second, it is impossible for any magnitude to arise from indivisibles, whether those be finite in number or infinite in number.

There are many evident demonstrations for both of these assertions. However, for the sake of brevity we shall point out only one or two, and we shall limit ourselves to a brief discussion, working crudely, reserving a more thorough geometric proof for some other time.

Against the existence of an infinite multitude.

Imagine an infinite multitude of individual items. Then imagine these individuals grouped into pairs.

Either the pairs themselves are infinite in number, or they are not. If they are not, then they are merely finite in number, and thus the individuals are also finite in number, contrary to the original statement.

If the pairs are infinite in number, then they can in turn be paired. These paired pairs, or foursomes, are now either infinite in number, or not. If not, then the pairs are merely finite, and so in turn the individuals are merely finite, contrary to the original statement. This can be extended to any grouping, any number of times.

Thus, if an infinite multitude can exist, it must divide into an infinite number of pairs, and an infinite number of threesomes, and an infinite number of foursomes, and so forth. An infinite multitude cannot divide to produce a finite number of groupings.

But if an infinite multitude must divide into an infinite number of pairs, threesomes, foursomes, and so forth, then an infinite multitude contains any number within itself, infinitely many times. This leads to one of two difficult possibilities regarding an infinite multitude.

One on hand, all infinites might be equal. Thus the infinitude of pairs in the multitude equals the infinitude of foursomes. But this is to equate the pair to the foursome—the part to the whole—which is impossible.

On the other hand, all infinites might not be equal; there might be different classes of infinite. Thus there is a greater infinitude of individuals than of pairs, a greater infinitude of pairs than of foursomes. But this is to say that one might be able to subtract an infinitude from an infinitude, with the difference being an infinitude, which is contrary to the concept of an infinitude.

Against an infinite, unbounded extension.

Imagine an infinite multitude of people standing side by side: A, B, C, D . . . K, and so on, as shown [in Figure 10-1]. In this group either there is a person who is separated from person A by an infinite distance, or there is not.

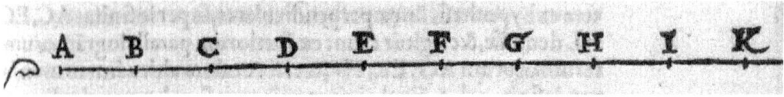

FIGURE 10-1

If there is a person who is separated from person A by an infinite distance, then suppose that person is Ψ (not shown in the figure). Then between person A and person Ψ there is an infinite interval, bounded on each end. We shall shortly demonstrate that this is impossible.

If there is not a person who is separated from person A by an infinite distance, then all points on the line AK, etc., are a finite distance from A. The line is a finite, not infinite, extension.

Another demonstration.

Two infinite straight lines extend out from A, these being AB and AC as shown [in Figure 10-2]. I say the distance BC contained between them is also infinite.

Indeed, suppose a perpendicular line EF, equal in length to interval D, is drawn upward from AC to AB, and then another infinite line EG is extended from E parallel to AC. Then from H (where EH is equal to AE) another infinite line HL is extended parallel to AC, and a line HI is extended perpendicular to AC. And, likewise, infinite line MP is extended from M (HM being equal to AE) parallel to AC, and MO is extended perpendicularly to it. And suppose this is repeated endlessly.

The created triangles AEF, EHK, HMN, etc., are identical, following Euclid (book 1, propositions 29 and 26),[49] and are of identical heights EF, HK, MN, etc. To the right of these triangles are unbounded rectangles of identical height, again following Euclid (book 6, definition 4).[50]

There are as many of these identical-height rectangles between AC and AB as there are equal line segments AE, EH, HM in line AB—namely, infinitely many. Thus between AC and AB is an infinite distance. I shall now demonstrate that this is not possible.

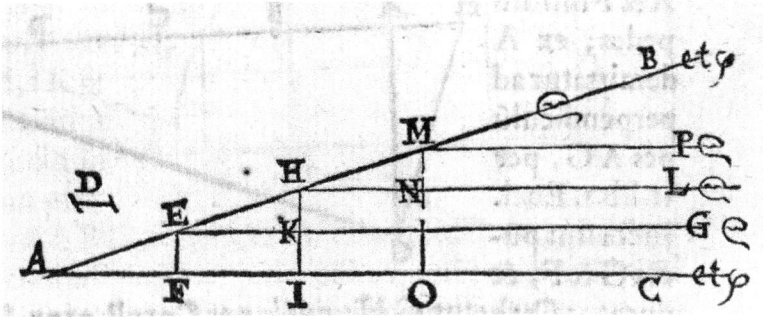

FIGURE 10-2

Demonstration against an infinity enclosed by boundaries.

Suppose that between A and F [in Figure 10-3] there are infinitely many feet. Suppose further that G is placed at a distance of one foot from A, measured perpendicularly to AF (Euclid book 1, proposition 11).[51] A line is drawn from G to F, and a line perpendicular to GF is drawn from G to H. Following Euclid (book 6, proposition 8),[52] the foot-length AG is proportionately in the middle between HA (which measures less than AG, or less than one foot) and AF (which exceeds AG). The ratio of HA to AG equals the ratio of AG to AF, or HA/AG = AG/AF.

However, the ratio of HA to AG is the ratio of a finite number to a finite number; HA/AG is finite. Therefore AG/AF is finite, and thus AF is finite, contrary to the original statement. Or, if AF/AG is the ratio of an infinite number to a finite number, then AG/HA is also, which is impossible because AG is one foot.

Therefore any actual infinity, whether closed by boundaries or not, is impossible.

Demonstration that it is impossible for a line to be constructed from indivisibles whether finite or infinite.

Note first: a line is essentially nothing other than continuous mere points. Second: points which are in the same place do not form a line. Third: one

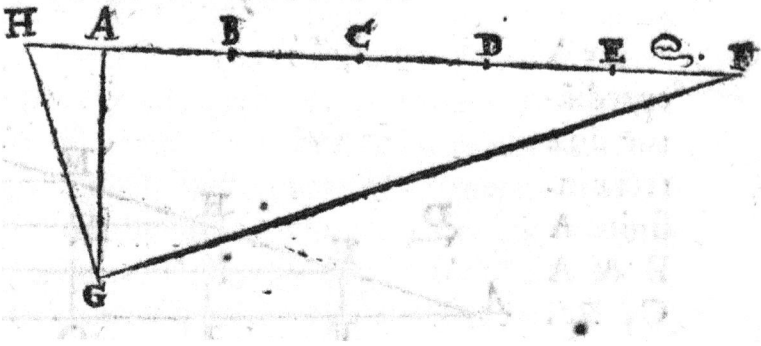

FIGURE 10-3

point is not larger than another point. Fourth: from the nature of the thing, nothing follows a point except another point. With this, we omit numerous demonstrations and instead touch on just one.

Consider a semicircular arc ABC, composed from mere points, as shown [in Figure 10-4]. From each point in ABC a parallel descends into the diameter AC. Illustrated in the figure are a limited number of these points (A, D, F, B, I, L, C) with their parallels descending into diameter AC (DE, FG, BH, IK, and LM). The parallels occupy as many points in arc ABC as in diameter AC. Since each unique point in ABC is connected by a unique parallel to a unique point in AC, there are equally many points in ABC as in AC. Thus, according to that axiom that says that equal numbers of identical things are identical, ABC and AC must be equal, which is impossible since, as Archimedes says, in fact the arc of a semicircle is not equal to the diameter.

Corollaries

1. Therefore, by reason of these, the universe cannot have existed from eternity.

2. Therefore, the universe was brought into existence at some point in time.

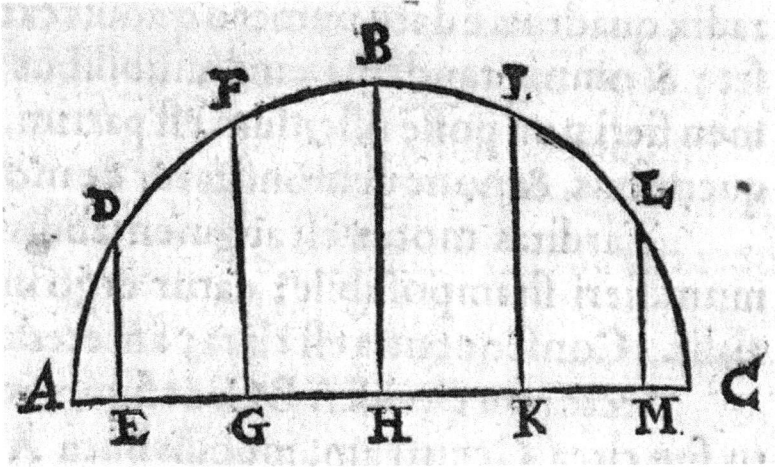

FIGURE 10-4

3. Therefore, we know by evidence that almighty God, the author of the universe, exists.

4. The fables of the ancients that describe the universe forming from the confluence of elementary particles are destroyed.

5. And contrary to Aristotle, the universe is neither inevitable nor eternal.

6. If an infinity closed by boundaries is granted, then it is evident that one such closed infinity may be greater than another. [In Figure 10-3] line GF (which subtends the right angle at A) will be greater than AF, following Euclid, book 1, proposition 19.[53]

DISQUISITION 11.

Just as we completely deny every actual infinity, we admit of necessity to potential or mathematical infinity—in number, in time, in motion and the quantities and qualities of motion such as rapidity or slowness, and so forth. Otherwise, irrational quantities might not exist in the nature of things, and circumferences could be calculated exactly in terms of diameters, the square root of every number could be determined exactly, and

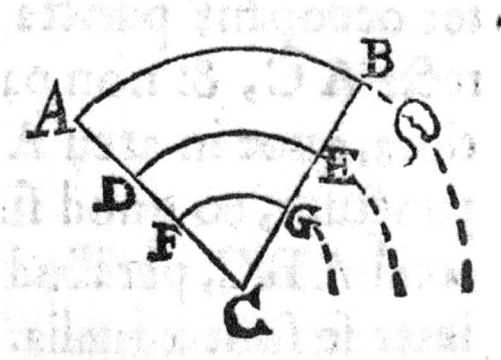

FIGURE 11-1

finally, all things might be composed of indivisibles (the impossibility of which was just demonstrated and will be further demonstrated later). Now we shall demonstrate this regarding slowness of motion.

It is clear that if there can be no slowest movement possible, then slowness of movement can be extended into infinity, and a certain potential infinity must be granted. Now, to demonstrate that there can be no slowest movement:

Imagine a sector ABC of a circle [as in Figure 11-1]. Line AC is made to rotate about the center C from A to B, such that as A moves into B, D moves into E, and F into G. Suppose the motion from A to B is deemed to be the slowest possible, or indivisibly slow. However, points D and F on line AC move through the same angle as does point A, and in equal time, yet they move through smaller arcs. Therefore, since points D and F move through smaller arcs in equal time, their motions are slower than the motion of A, contrary to the statement that the motion from A to B is the slowest possible. Thus in slowness a certain infinity does exist, which is what we set out to prove.

Concerning the different parts of the universe

Earth and Water

DISQUISITION 12.

Earth stands at rest in the center of the universe. It is not a planet like Mars. It is subject to neither annual nor diurnal motion. Rather, the rest of the universe revolves about Earth in a circle, while Earth, the center of the universe, is held motionless by its own heaviness. Many ancient philosophers and mathematicians have opposed this truth, and many more recent ones, if not the majority, oppose it now (since now the majority is busy burning incense on the altar of Copernicus). But we follow reasoning and experience, and we are by no means easily swayed by assertions.

DISQUISITION 13.

Copernicus, that otherwise most learned man, has taken Earth and the whole world of the four elements, together with the moon, and raised it all up to a place between the circles of Mars and Venus.[54] He has driven the sun down into the center of the universe. He has made the sun and the firmament of stars fixed. He has made Earth mobile, with a triple motion: annual [around the sun yearly], diurnal [around its own axis daily], and a third motion [that maintains the axis of Earth in the direction of the North Star].[55]

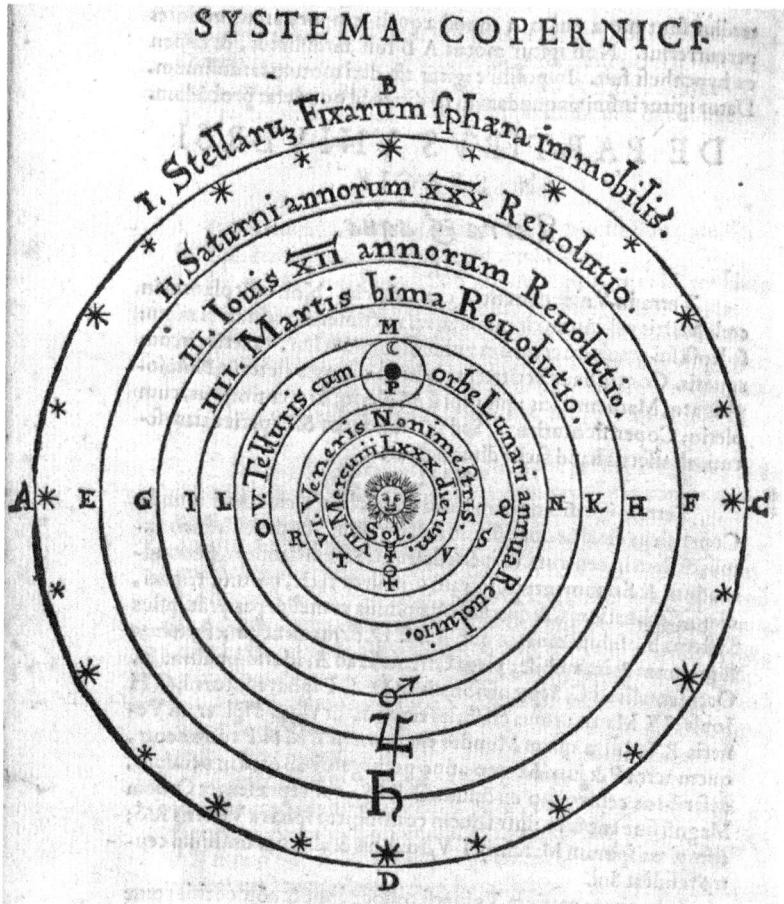

FIGURE 13-1

This system is illustrated here [in Figure 13-1]. The firmament of fixed stars ABCD is the highest celestial circle and is immobile. East is toward A, south toward B, west toward C, north toward D. The circle of Saturn is EF. The circle of Jupiter is GH. The circle of Mars is IK. Within it and extending down to the circle of Venus RS is the region of the world of the elements LMNP, including the moon M and Earth P. Earth, with the moon, moves from east to west, with the center of Earth describing the circle OPQ—the Annual Circle, the Great Orb. Within all this lies the circle of Venus RS, and within that is the circle of Mercury TV. The sun resides in the center of all.

In this hypothesis—because of the arrangement of the parts of the universe and the imagined motion of Earth—the sun, Mercury, and Venus are below, and Earth is above. Heavy bodies absolutely and naturally ascend. Light bodies descend. Christ the Lord ascended to hell.[56] He descended to heaven (for he approached[57] the sun). When Joshua commanded the sun to stand still, actually Earth stood still, or the sun moved contrarily to Earth. Summer occurs not when the sun is in the summer constellation of Cancer, but when Earth is in the winter constellation of Capricorn, and vice versa for winter. The stars do not rise and set to Earth, but rather Earth to the stars—rising becomes setting, setting becomes rising.[58]

In short, the whole course of the universe, as it were, is inverted. To this objection, and to others like it, the Copernicans can provide answers which are satisfactory, if torturous. However, they will be less able to satisfy the objections which we shall now present.

An argument based on Earth's annual motion around the sun.

In the opinion of Copernicus and of all the Copernicans, the semidiameter of the circle of Earth's orbit around the sun is of negligible size compared to the distance to the fixed stars. ⟨This can be seen in Copernicus's *On the Revolutions*, book 1, chapter 6 and elsewhere.⟩[59] However, this is absurd, and it gives birth to other absurdities, as we shall now show.

According to the Copernicans, the distance between Earth and the sun—that is, the semidiameter of the orbital circle—is 1,208 semidiameters of Earth, or 1,208 terrestrial semidiameters. Following the Copernican opinion that the semidiameter of Earth measures 860 common German miles, this is 1,038,880 German miles. ⟨See page 168 of Maestlin's appendix to Rheticus's *First Account Concerning Copernicus*.⟩[60] But the sun has a measurable diurnal parallax.[61] This is well known, granted by Copernicus himself and by his minions.[62] Thus, 1 part in 1,208 is not negligible.[63]

Nevertheless, let us suppose the distance to a fixed star to be 1,208 semidiameters of the orbital circle, so the annual parallax of that fixed star is the same as the diurnal parallax of the sun. Thus the distance to that fixed star is 1,208×1,208 = 1,459,264 terrestrial semidiameters, or 1,254,967,040 German miles.

However, as remote as that may be, the star is still subject to annual parallax. The distance is still not great enough for the size of the orbit circle of Earth to be negligible by comparison. Thus the star must be yet more distant.

Saturn is more than nine times as distant from Earth as is the sun. Saturn is therefore distant from Earth by more than 9×1,208 = 10,872 terrestrial semidiameters, or 11,294,703,360 German miles. Saturn shows no diurnal parallax, so 1 part in 10,872 is indeed negligible. Thus, let us suppose a fixed star to be distant by 10,872 semidiameters of Earth's orbital circle. At such a distance we expect to see no annual parallax. This star is distant by 1,208×10,872 = 13,133,376 semidiameters of Earth.

Now consider that the diameter of the sun is 11 semidiameters of Earth,[64] while its apparent diameter seen from Earth is one-half of a degree, or 30 minutes of arc,[65] and its distance from Earth is 1,208 terrestrial semidiameters. A bright star such as Regulus has an apparent diameter of 2 minutes of arc, which is fifteen times smaller than that of the sun. We take the distance of Regulus to be about 13,133,376 terrestrial semidiameters, as noted above, or 10,872 times more distant than the sun. Thus, were the sun at the same distance as Regulus, yet measuring 30 minutes in diameter, it would be 10,872 times larger than it truly is, or 10,872×11 = 119,592 terrestrial semidiameters. But it would also then be fifteen times larger than Regulus in terms of true size. Thus the true diameter of Regulus is 119,592/15 = 7,972 terrestrial semidiameters (ignoring fractional amounts).

Thus any fixed star in the firmament whose apparent diameter measures two minutes of arc has a true diameter of 7,972 semidiameters of Earth. A fixed star whose apparent diameter is one minute of arc has a true diameter of 3,986 semidiameters of Earth. According to Tycho ‹(book 1 of his *Introductory Exercises*, page 481)›, the apparent diameters of stars of the first, second, third, fourth, fifth, and sixth magnitude are two minutes, three-halves of a minute, a little over one minute, three-quarters of a minute, half of a minute, and one-third of a minute, respectively.[66] Thus, as a first-magnitude star has a true diameter of 7,972 terrestrial semidiameters, and since the diameter of the circle of Earth's orbit contains 2×1,208 = 2,416 terrestrial semidiameters, a first-magnitude star has a true diameter of 7,972/2,416 = 3 & 181/604 times the diameter of Earth's orbit. By way of the same calculations, a third-magnitude star has a true diameter that is 1 & 785/1,208 times the diameter of Earth's orbit, and a sixth-magnitude star has a true diameter slightly more than half the orbital diameter. In other words, a first-magnitude star would reach from Earth out past Mars toward Jupiter.[67] Even a small sixth-magnitude star would reach from Earth to the sun.

And in terms of volume, since according to Euclid (book 12, proposition 18)[68] volume proceeds as the cube of diameter, *a first-magnitude star would be more than thirty-five times greater than the spherical volume enclosed by Earth's entire orbit.* And so, following the Copernican opinion, the volume of the star Procyon, the Little Dog, is *over thirty times greater* than the sphere enclosed by the orbit of Earth.

The number 13,133,376—determined using the ratio of the size of the orbit of Saturn to the size of Earth—is not excessive. Tycho Brahe, in book 1 of his *Introductory Exercises*, pages 480–81,[69] determines the distance of Saturn from Earth in the Copernican opinion to be 12,900 terrestrial semidiameters, and says that the firmament of fixed stars would need to be more than 700 times more distant than that. So, if we use 800 as the distance value, we obtain by way of multiplication 800×12,900 = 10,320,000 terrestrial semidiameters. This distance will yield almost the same sizes for the fixed stars of the firmament that we calculated previously.

Copernicus himself ‹in book 1, chapter 6 of *On the Revolutions*› has given heed to this amazing immensity of the firmament and vastness of the fixed stars.[70] His minions do not deny any of this ‹(see Rheticus's *First Account*, page 118)›.[71] Instead they go on about how from this everyone may better perceive the majesty of the Creator.[72] This is laughable, since the stars appear so small, and even the most learned person cannot easily perceive this monstrous size.

However, both in the past and today, the wiser astronomers have been rightly vexed at this utterly vacuous monstrosity. Thus ‹in *The Sand Reckoner*› Archimedes has said,

> Aristarchus of Samos brought out a book consisting of some hypotheses, in which the premisses lead to the result that the universe is many times

greater than that now so called. His hypotheses are that the fixed stars and the sun remain unmoved, that the earth revolves about the sun in the circumference of a circle, the sun lying in the middle of the orbit, and that the sphere of the fixed stars, situated about the same center as the sun, is so great that the circle in which he supposes the earth to revolve bears such a proportion to the distance of the fixed stars as the centre of the sphere bears to its surface. *Now it is easy to see that this is impossible.* . . .[73]

From this it follows that even the smallest star visible to the eye is much larger than the whole circle of Earth's orbit. This is because such a star has a measurable size, as does the circumference of the sky. The ratio of the size of the star to the size of the firmament of fixed stars is therefore perceptible. But according to the Copernican opinion, the ratio of the size of the circle of Earth's orbit to the size of the firmament is imperceptible. For in the Copernican opinion the size of Earth's orbital circle holds the same proportion to the firmament as the size of Earth itself holds to the firmament in the common geocentric opinion. Yet experience shows Earth to be of imperceptible size compared to the firmament.[74] Thus, in the Copernican opinion it is the circle of Earth's orbit that is of imperceptible size compared to the firmament—and therefore smaller than the smallest perceptible star.

All of which leads to three questions. First, what good are such enormous contrivances, set around this tiny point of Earth? Second, why are they so far removed, so that they appear so trifling and hardly able to shine on Earth? Third, what is the point of that insane chasm between the stars and Saturn? All these things are certainly to no purpose. No plausible reasoning can defend them.

Thus ‹in book 1 of his *Introductory Exercises*, page 685› Tycho Brahe has written well:

Copernicus has preferred rather to grant another sort of absurdity, no less inharmonious and incredible, a vastness of such size evidently to be encompassed between most distant Saturn and the sphere of the fixed stars, that with respect to it the annual orb of earth might entirely vanish, rendered insensible because of that excessive gap, that is, so that no proportion might exist. *About this thing, how it may be beyond all belief, and how many absurdities may follow this arrangement, we shall discuss elsewhere.*[75]

The argument in a nutshell.

1. The Copernican motion of Earth requires both many passages of the sacred Scripture, and many common ways astronomers speak, to be twisted in a preposterous sense—and without any need, since the plain sense of all can be defended most easily. Therefore his opinion need not be accepted.

2. The Copernican motion of Earth makes any first-magnitude star thirty times greater than the volume enclosed by the circle of Earth's orbit, and makes any little star seen in the starry heaven greater than that volume. Therefore his opinion need not be accepted.

3. The Copernican motion of Earth introduces between Saturn and the fixed stars a certain immense desolation of distance, the use of which Copernicans themselves do not know. Therefore his opinion need not be accepted.

DISQUISITION 14.

An argument based on Earth's diurnal rotation.

The annual motion of Earth requires the Copernicans to assert a daily rotation of Earth. Otherwise, one hemisphere would continually face toward the sun, and the other would continually be dark. But we will now demonstrate this gyration of Earth to be impossible.[76]

Everyone, even Copernicans, acknowledges heavy bodies to be pulled down toward the center of Earth along a vertical line, and light bodies to be borne up from the same center along the same vertical line. ‹See Kepler's *Optics,* page 309, or Simon Stevin's definitions 2 and 5 on statics from his *Mathematical Notes.*›[77] Thus [in Figure 14-1], the little ball A, the snail E, the hail, snow, and rain G, the seagull M, the plumb line TV, and the pebble X tend straight down toward the center of Earth C. Conversely, the smoke K from the chimney L, the arrow R, the hurled cannonball Q, and the rocket S all tend up from the same Earth BNL. Meanwhile the sweetly warbling lark soars above its little nest O, according to perpendicular OP; the raven D is above the rocks F; and the seagull M is over the little fish N. Indeed, such things occur daily.[78]

Now, *it is an undeniable fact of the Copernican opinion* that Earth turns with a twenty-four-hour period. Thus, any point on the equator is

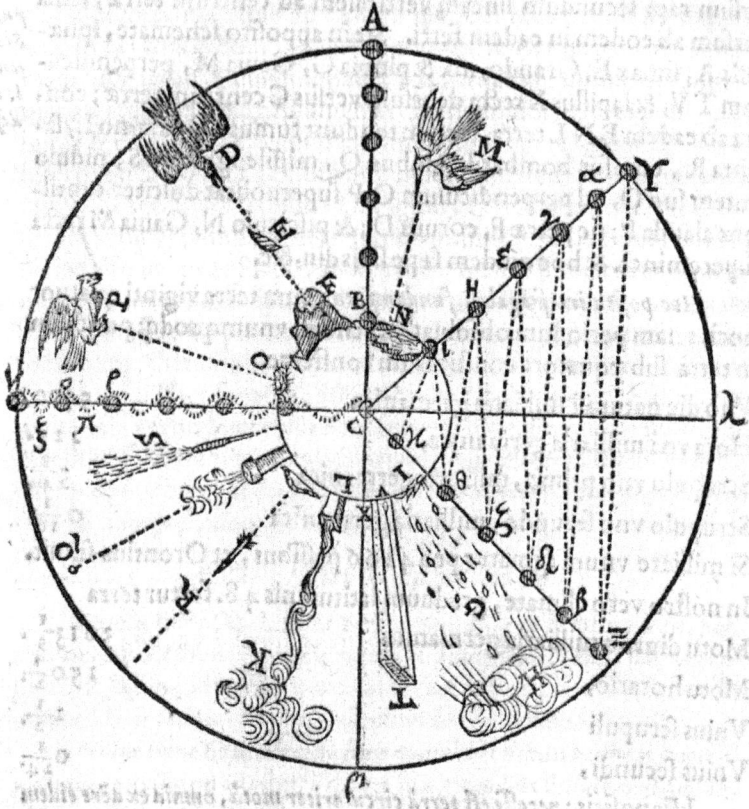

FIGURE 14-1

carried 5,400 German miles in one day (one mile being 4,166 paces), 225
German miles in one hour, 3 & 3/4 German miles in one minute, and
one-sixteenth of a German mile in one second.[79] At our latitude of 48 de-
grees, Earth's diurnal rotation carries any point 3,613 & 1/3 German miles
in one day, 150 & 1/2 in one hour, 2 & 1/2 in one minute, and one twenty-
fourth of a mile in one second.[80]

Granted this, *it is necessary on a rotating Earth that all things ap-
proaching Earth perpendicularly out of the air be carried along with Earth
by a motion proportional in all manners.* Consequently, the small balls A
(above point B), ♈ (above point ι), and ν (above the pole of Earth),
will all be carried around with the same period of rotation as Earth.

Earth BNX [in Figure 14-1] rotates on axis νλ—from B through X and back to B.

Now, let us specify these three balls to be equal in weight and size. Let us further imagine that they are placed at a distance from Earth equal to that of the moon, and allowed to freely descend. Finally, let us suppose the motion downward is equal in terms of swiftness to the moon's orbital motion. Then a minimum of six days will elapse as a ball descends.[81] (This is different from how, in the Copernican opinion, ball A at the distance of the moon will be caused to circle in one hour through 12,600 German miles, and ball Υ at our latitude through 8,431 German miles [in order to remain above the same point on Earth]. These quantities are determined according to Copernicus and his follower Maestlin ‹see *On the Revolutions*, book 4, chapter 17 and Maestlin's appendix to the *First Account*, page 167[82]›. It is not possible that the straight descent should be equal in speed to such an onrush—we can grant this much to our adversaries.)

During these six days ball A will circle six times around Earth. It will circle with Earth,[83] in the free air, so that it always remains above its point B. Ball Υ will also circle six times. Twelve hours after it is released it will arrive at Z; twelve hours after that (twenty-four hours after it is released) it will arrive at α. In twelve more hours it will be at β, and from there another twelve takes it into γ. Thus it will keep turning round proportionally with Earth, from γ through δ into ε; from ε through ζ into H; toward θ, toward ι, toward κ; into C the center of the world. The same will happen in similar proportion, with the birds, the smoke, the fire, the clouds, and the other things that steadily remain in the free air over one place for some span of time.[84]

All this is firmly established by ongoing experience. Therefore some questions arise that require answers from the Copernicans.

Questions.[85]

The primary question is, *what is the origin of that circular movement of heavy and light objects? Is the origin internal or external?*

Questions if the origin is external.

- Is it God that drives the circular movement, through a continuous miracle? Is it an angel? Is it the air?

Indeed, it is to the air that many attribute the origin of the circular movement. But consider the following questions:

- What revolves the air around? *Is it a natural motion, or is it a forced motion?* The former is contrary to truth, contrary to experience, and contrary to Copernicus (book 1, chapter 8).[86] The latter will not suffice, as the following questions show.
  - How might it happen that the air, being pulled along forcibly, follows Earth everywhere, *matching its motion inviolably?*[87]
  - Why is the resistance of air in this case either absent or not apparent, *contrary to every experience of the force of its flow?*
  - What cause would explain how the force of contact from Earth that causes the air to move with Earth is weakest *near Earth* and strongest *far from Earth?* If the air is what hauls the heavy bodies within it into circular motion, then the speed of the air must necessarily be greater at greater heights above Earth.[88] And so, therefore, the force that Earth exerts to drag the air along must be applied most weakly to the air at Earth's surface and most strongly to the air scattered high above. How can this be? Experience and reason both cry out in protest.
  - How does the air, dragged along by Earth, in turn drag other bodies, both light and heavy, along with it, *in exact proportion to the terrestrial revolution?*
    - And what trick allows that air to equally drag away *a plume of feather and a plumb line's leaden bob? Smoke rising and shit dropping?*[89]
    - Why can't the strongest winds (of a kind that can topple armored knights off their horses) drag plume and plumb bob, smoke and shit, equally?
    - Why is that ability to equally drag plume, plumb bob, smoke, and shit not present in water, an element *denser* than air? Pebbles carried in torrents of water are not dragged equally by that water. Rather, the heavier pebbles tend to the bottom and resist the action of the water. Experiences of this occur everywhere. For example, a rock dropped from a ship floating on a river will not be carried along by the current as rapidly as the ship, but as the rock settles to the bottom it will fall behind.

And denser fluids are more suitable to dragging than are more tenuous ones. But in the Copernican opinion the contrary is true: the same pebble, which certainly resists water's dragging, offers no such resistance to air.

- Finally, the higher region of the air is at rest according to Copernicus ‹(book 1, chapter 8)›.[90] What of that?

Through these questions and others like them, it is clear that by no means can this motion of heavy and light bodies be of external origin. But if the origin of the circular movement of heavy and light objects, in the Copernican opinion, is internal, according to nature, then the most difficult questions—in fact, inextricable questions—arise.

Questions if the origin is internal.

- That supposed internal origin is either some nonessential property the object happens to possess, or it is inherent to the very substance of the object itself. If it is some nonessential property, then what kind of property? Until now, no "moves in a circle about a point" property has ever been recognized. Although such a property might exist, how might it be found in such contrary things? *In fire as in water? In air as in earth? In living things as in inanimate objects?* If it is inherent, arising from the very material nature of the object, then what of the diverse natures of different objects? That very diversity disagrees with this proposition. Birds, snails, stones, arrows, snows, smokes, hailstones, fish, and so on—all are different in many ways: *how could it be that all move circularly on account of nature, even though their natures are so greatly diverse?*[91]
- Suppose God willed Earth [in Figure 14-1] to cease rotating. Would the other objects continue their circular motions, or not? If not, then circular motion is not a product of their natures. If so, then the above questions return. And it certainly would be remarkable how the seagull M could not remain over the little fish N, nor the lark P over its nest O, nor the raven D over the snail E and rocks F, even though each would wish to. Yet it is clear that, should that circular motion which would pull them along be simply natural, *an animal could not and would not be able to offer resistance.* Moreover, the animal would be carried along the same, were it dead or alive.

- In addition, why are things so different moved only from west to east, parallel to the equator? And moved continuously, never stopping?
- By what means are those things which are higher moved faster, and those which are lower moved slower,[92] as is seen in ʾΥ, Z, and α, β, and so on [in Figure 14-1]?
- By what means are those things closer to the equator moved faster (through larger circles) and those farther from the equator moved slower (through smaller circles), as is seen in A and ʾΥ?
- How does it happen that a ball at A, on the equator, is revolved around the center of Earth C, through a great circle, at incredible speed—and yet at the pole ν it is revolved around its own center, through no circle at all, and very slowly? (Were the ball descending, then in one day its edge would travel from ν into ξ, and into o during the descent.)
   ○ Should not the same ball, if it revolves around the center of Earth through a great circle when at A, also revolve around the center of Earth through a great circle everywhere? But instead, when it is moved away from the equator, it goes through a smaller circle.
- If circular motion is natural to light and heavy bodies, then straight motion is of what sort? If it is natural, how? *For the circular motion is natural, and straight motion is of a different kind.* If it is forced motion, why does it happen that as the rocket S [in Figure 14-1] is rushing up, with its head spewing sparks, it does not curve, *since the upward movement may be weaker and forced, and the circular movement more vigorous and natural?* Why does the rocket leave a trail of falling sparks that is perpendicular to the ground?[93]
- How can it be that, when spheres A, ʾΥ, and ν each fall, the center of A describes *a spiral in a single plane,* while the center of ʾΥ traces out *a spiral line on a cone,* and the center of ν traces *a straight line coincident with its axis* while a point on its edge traces out a spiral on a cylindrical surface?

Copernicus has said ‹(in *On the Revolutions,* book 1, chapter 8)› [that the straight motion of heavy and light bodies down and up is the motion of bodies that are not in their proper condition—of bodies that are out of their naturally ordered place, away from the whole to which they belong. They move straight so as to reach their proper place and return to the whole. But, says Copernicus, circular motion goes on continually. Thus, he says,] *circular motion endures through the straight motion of heavy and light bodies, like life endures through sickness.*[94] Most of his minions endorse this and expressly state that straight motion exists only so that heavy bodies may move down to join to the rest of the whole heavy Earth below, and light bodies may move up to join with the rest of the realm of fire above. And so, in their opinion, were the whole Earth [in Figure 14-1] to disappear, *then no hail or rain G from cloud H would fall, but rather it would be carried naturally in a circle; neither would fire nor anything fiery ascend.* Although in their opinion there probably is no fire above (something reason and experience oppose, and Copernicans cannot escape this by invoking density, for enlarging something so as to make it less dense does not make it less heavy, and will not make it rise violently like fire).

Now imagine Earth to be hollow and full of air [as in Figure 14-2]. A stone G placed at the center of Earth A either will ascend to the surrounding earthy shell, at some point C, or will not. If it will not, then it is not true that parts move so as to be joined to the whole. If it will, *then heavy bodies do not rest in the center by reason of their own weight—all reason and experience rebels against this.*

Likewise, if a stone F is released from the concave part B of this hollow Earth HI, either it will move into the center A, or it will not. If so, then it separates itself from the whole, against Copernicus; if not, it opposes all experience, since we see entire arches fall in, among other examples.

Therefore straight motion is not something that happens to heavy and light bodies, as the Copernicans wish, but rather it is of the nature of those bodies. Their supposed circular motion cannot be natural motion and cannot be forced motion, as we have shown. Therefore, it cannot be. Therefore, it is impossible for Earth to go in a circle. *Therefore, in brief: the hypothesis of the motion of Earth is fictitious; the hypothesis of the motionlessness of Earth is true.* Goodbye, then, to Earth being one of the planets.

There is also a certain particular defender of Copernicus who says that Earth rolls around the sun like a wheel, so that its daily revolution generates its annual orbit. He does not see that *this would make either the annual orbit too small or the globe of Earth too large,* contrary to the opinion of Copernicus himself, and contrary to true knowledge. For, following the opinion of that author, the great orbital circle that Earth traces out in one

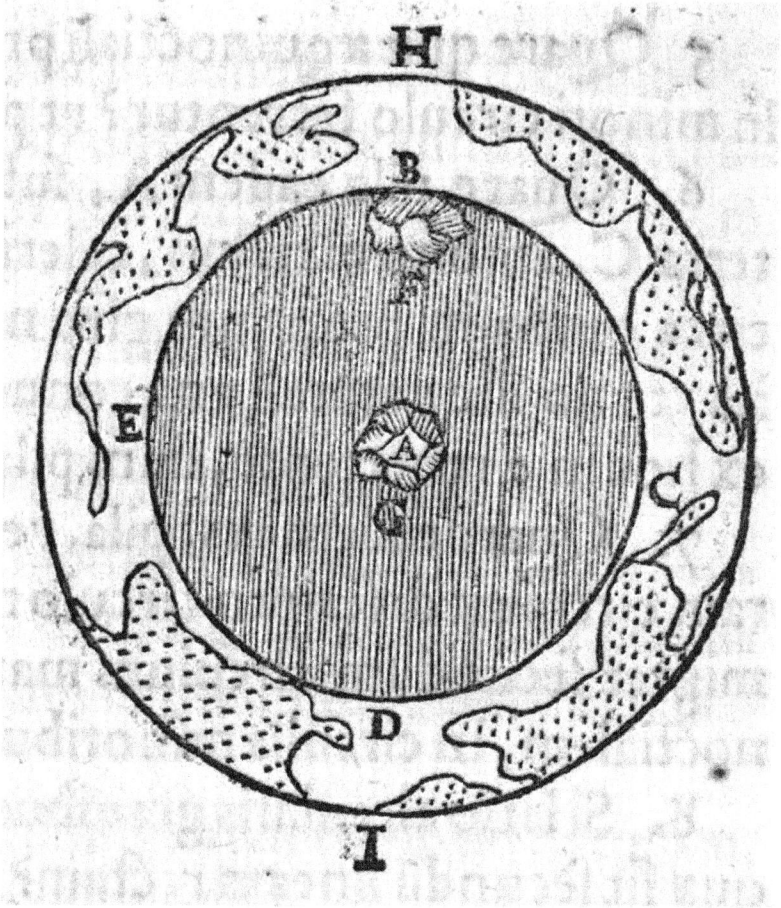

FIGURE 14-2

year is equal to 365 circumferences of the terrestrial equator, for the reason that the terrestrial globe may pay out as many turns as there are days in a year, namely 365. Now, according to Copernicus, the semidiameter of Earth is 860 German miles, the diameter 1,720. The diameter tripled will give the approximate full circumference of Earth, 5,160 miles.[95] Earth runs through this distance on its orbital circle each day (following that author). Therefore, since Earth in one year pays out the whole orbital circle, composed of 365 circumferences, each of which is 5,160 miles (corresponding to the distance travelled in one day), the circumference of the orbital circle will be 365×5,160 = 1,883,400 German miles. This is too small and contrary to Copernicus and all others, who say the semidiameter of the orbital circle to be 1,038,880 miles, the diameter 2,077,760, and consequently the circumference 3×2,077,760 = 6,233,280. This number is more than triple 1,883,400. But if you contend the Copernican value for the orbital circumference, Earth now becomes too large. Indeed, if you divide the number 6,233,280 by 365, it will yield as the quotient 17,074,[96] which will be the circumference of Earth in miles. This again is against Copernicus and all others, since this number is more than triple 5,160.

The argument in summary.

1. The Copernican motion of Earth takes away from the universe the simple motion of things up and down. Therefore the motion of Earth need not be granted.

2. The Copernican motion of Earth necessarily introduces a certain circular movement to all things moving up or down. Regardless of whether that motion is said to be natural or not natural, internal or external, innumerable absurdities arise on account of it. Therefore the motion of Earth need not be granted.

3. The Copernican motion of Earth makes straight motion by heavy bodies something not so much natural as almost forced, and makes circular motion necessary and natural. It gives a single natural motion to things most diverse by their natures. It summons stones up from the center of Earth so that it can drop them down from arches, while they turn with or move toward the whole by means of nothing more than love. All these things are against reasoning and experience. Therefore the motion of Earth need not be granted.

DISQUISITION 15.

The center of the universe is the place where elemental earth rests. All straight lines drawn from there to the outer sphere of the heavens are of equal length. This is on account of the heaviness of elemental earth, as Macrobius elegantly explains in book 1, chapter 22 of his *On Scipio's Dream.*

Nature always draws weights towards the bottom; obviously this was done that there might be an earth in the universe. Of all the matter that went into the creation of the universe, that which was purest and clearest took the highest position and was called ether; the part that was less pure and had some slight weight became air and held second place; next came that part which was indeed still clear but which the sense of touch demonstrates to be corporeal, that which formed the bodies of water; lastly, as a result of the downward rush of matter, there was that vast, impenetrable solid, the dregs and off-scourings of the purified elements, which had settled to the bottom, plunged in continual and oppressing chill, relegated to the last position in the universe, far from the sun. Because this became so hardened it received the name *terra*.[97]

A single globe arises from the water and the earth. The geometric center of this globe, and the center of gravity of the globe, are one and the same ‹(see Clavius on *The Sphere*, tome 3, chapter 1).›[98]

The center of gravity is that point within a heavy body that would occupy the center of the universe, were that body able to freely move to, and naturally come to rest at, the center of the universe ‹(see Clavius again; and Simon Stevin, volume 2, book 1 of *On Statics*, definition 4)›.[99] Thus a single heavy body has one single center of gravity—one point at which it would rest, were it at the center of the universe. That point would be changed, were there to be additions to or subtractions from the body. Here a "single body" is defined loosely: one body may be composed of separate bodies packed and bonded together.

From these I deduce the following conclusions.

Conclusions.

1. Since additions to and subtractions from the terrestrial globe occur continually (through the generation of some things and the eroding away of others, through evaporations and eruptions, etc.), its center of gravity is necessarily changing.

2. Because of this the globe itself must vacillate with a certain perpetual, but entirely insensible, trembling.

3. Newly altered parts of Earth apply their weight continually until the center of gravity of Earth coincides with the center of the universe.

Were this not the case, then Aristotle's whole reasoning on why the center of Earth coincides with the center of the universe would collapse.

But it is uncertain whether the parts of Earth continue to apply their weight once the center of gravity of Earth is united with the center of the universe. Many things suggest they do; many things suggest they do not. We remain in the middle, unpersuaded either way.

4. *Artificial perpetual motion is not incompatible with nature.* ‹Such motion is possible.› Imagine a certain heavy gnomon ABC [Figure 15-1] which can pivot about A, the center of the universe. It is bound to axle DE, supported by stanchions DF and EG, and rotates about poles D and E. These granted, I say the following will occur: when gnomon ABC is rotated from C into H toward I, because of this action it both returns into C and continues on from there into H again, as before, and so on into perpetuity.

The cause of this continuing motion is forced suspension. For the whole gnomon is weightier toward its part C, on account of its shape; even more so if an iron ball S is appended to the gnomon at C. Vertex A is like the unyielding perch of a scale, for the axle supported at D and E may not move from the center of the universe. Thus all the points of the sphere S and of the gnomon ABC continually press down toward A, but because line BA is fixed at A, they do not fall into the center. They are impeded.

Thus their propensity to fall causes a force to be applied on BA. They haul down on it and urge it to incline, which it does not do on account of the rigidity of the gnomon. Thus all the pushing pours onto the pivot point A on the axle, or onto the rotating poles of the axle D and E. These, being free in their holes, allow nature to take its course. Thus by no means is perpetual circular motion hindered. That these things happen this way is confirmed by reasoning and manifested though common experience in statics.

‹One method of reasoning is to› imagine curve MN to be the surface of Earth and imagine the gnomon to be cut off there at MN, so that it stands upon that surface. It will topple in the direction of C and N, on account of the heavy portion MKC. This is obvious by common experience. The same happens if curve OP or curve QR is imagined to be the surface of Earth: the gnomon always topples toward C. But if the gnomon is intact (not cut off), the force exerted at N carries over into BA and in the end is directed toward the center. ‹Another method of reasoning is to› imagine a second identical gnomon attached against this one, whose weight is in the opposite direction. Then everything would stand in equilibrium and no

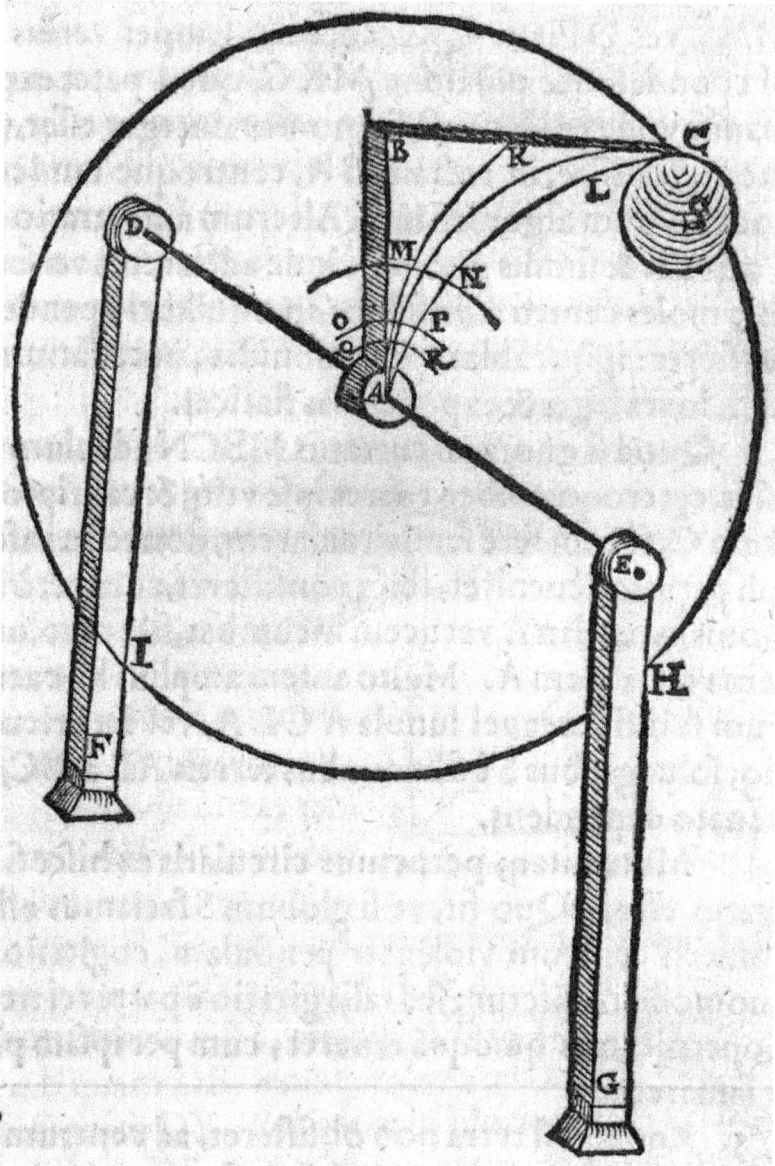

FIGURE 15-1

motion would occur. But remove the one half, and the other must move, as is experienced in scales and in statics.

Were the cut-off gnomon MBCN bound to the single point N, and otherwise free to move, point C would certainly fall, and while falling it would describe a generally semicircular arc until it had reached its lowest point, and there it would stop. (This would happen all the more were that gnomon attached to the center A shaped like a crescent, like ACLA, or like a hatchet, like AKC; or if it simply consisted of globe S attached to a bent iron rod AB-BC, or to a single iron arc ANC.) Now surely in the situation of the intact gnomon (not cut off), when the entire force presses into vertex A, a complete and perpetual revolution around that same A will occur.

And so, by reason of the preceding, we have demonstrated that perpetual circular motion is possible. Thus if we make globe S to be Earth, in forced suspension around the center of the universe A, we may be able now to imagine a manner by which Earth might be made to revolve around that center.[100] But such an orbital motion does not exist, and if it does exist, it is of no help to the Copernicans, since no observations are explained by means of it.

5. If not obstructed by earthen material, water will naturally flow toward the center of the universe, which is also the center of gravity of water, and the center of Earth, as Clavius shows ‹in tome 3, chapter 1 of his commentary on Sacro Bosco's *Sphere*›.[101] Thus the heaviness of the elements of water and earth differ not by kind, but by degree.

DISQUISITION 16.

Water by nature is lighter than earth, and floats upon earth, and with earth makes a single globe, in which the center of gravity, the geometric center, and the center of the universe are one and the same. But this globe, Earth, is an approximate globe, not a geometrically perfect globe. For according to Clavius ‹again in *Sphere*, tome 3, chapter 1›, the greatest height of a mountain, and the deepest depth of the seas and valleys, is about one German mile. Therefore each of these is, in comparison to the semidiameter of Earth, as 1 part to 860.[102] If we consider a perpendicular distance of two German miles from the deepest depth to the highest peak, then this will be as 1 part to 400, approximately. That one part is hardly sensible among the 400. Indeed, who could sense a single gold coin missing from 400? How much less sensible is that same distance in comparison to the distance from Earth to the planets, or to the fixed stars? Thus those who persuade themselves that the stars appear larger when viewed from a high mountain, and smaller when viewed from a deep valley, are mad. Even the semidiameter of Earth itself causes no sensible variation in the apparent sizes of stars (as is obvious, since the size of a star does not alter from rising, to transiting, to setting). How much less variation will result from two German miles?

Consider [Figure 16-1]. A is the center of Earth. Arc BC is the diurnal motion of the sun over six hours. AB is a vertical line. DE is a line to the visible horizon, as seen by the eye of an observer located on the surface of Earth at D.

I say D to be less distant from the vertical point B than from the horizontal point E, by an amount equal to the semidiameter of Earth DA. For DE is parallel to the horizontal line AC. Arc EC is approximately a straight line, and angle ACE is approximately equal to angle CAD. Therefore DE and AC are equal. AB is equal to AC, so AB and DE are equal. AB exceeds DB by the semidiameter of Earth DA, and so DE exceeds DB by the same amount.

When the sun is on the horizon at E, it is seen through distance DE, which is greater than the distance DB, through which the sun is seen when at noon. And since the sun does not appear smaller at E than at B, it is obvious that the semidiameter of Earth, which is FE, vanishes by comparison to DE. How much less perceptible will be a distance of two miles closer to or farther from the sun?

I said DE is equal to AC through approximations. In reality DE is less than AC, according to Euclid, book 3, proposition 7.[103] Accordingly, as DB and DF are equal, DE will exceed DB by less than the semidiameter of Earth, DA. That difference, however, is insensible.

In like manner, when the sun is at H it is farther from D, by the amount GH, than when it is at B (DG and DB being equal semidiameters of the same circle). From this truth some conclude wrongly that the sun shines hotter at noon than at sunrise or sunset, because at noon it is closer by a whole terrestrial semidiameter. But this is nonsense. ‹Such opinions

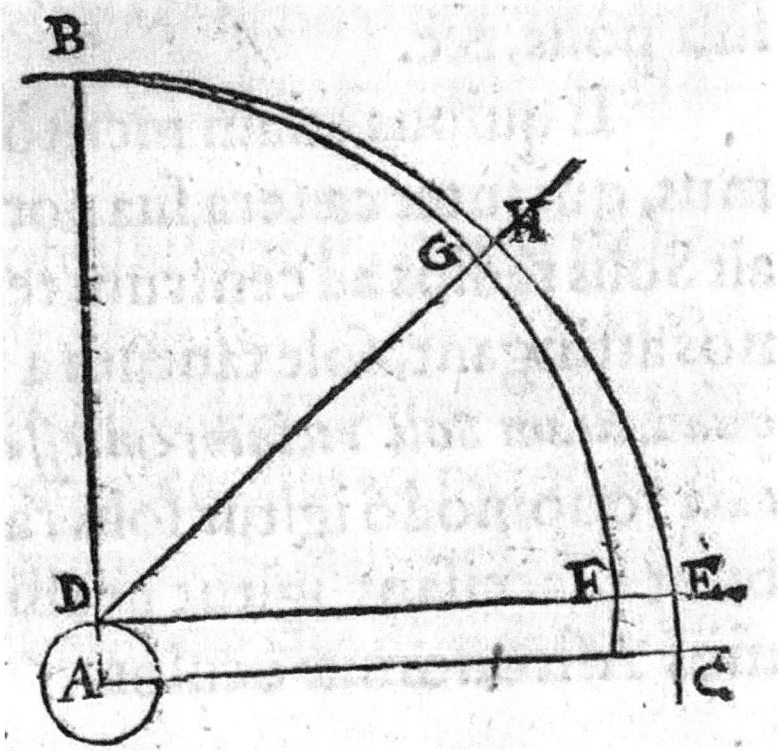

FIGURE 16-1

are to be rejected.› In winter the sun is not one but forty terrestrial semidiameters closer to Earth than in summer, as all astronomers will say,[104] and so according to this manner of thinking the sun ought to shine all the hotter in winter than in summer, which is the opposite of what happens. ‹Why then does the sun shine more or less hot?› *Because it is the direction or angle at which the solar rays impinge upon the ground that causes the strongest heat.* And so we see clearly the reason why the sun shines hotter at noon than in the morning or evening, hotter in the summer than in the winter, hotter within the tropical region than outside the tropics (and why the sun heats the poles so weakly), and so forth.

These things blunt the assertions of a certain astronomer, who among other things denies that the moon is illuminated by the sun, because, he says, the sun's rays only reach to the center of Earth. How might that be true, since those solar rays strike us in the summer, when the sun is so much more distant from us? Not to mention that *the quarter moon is closer to the sun than is Earth,*[105] and *the crescent moon is much closer.* How is it that the rays of the sun would strike Earth, then, yet not the moon? May deceptions of this sort go away, and not bewitch the purest eyes of astronomy.

DISQUISITION 17.

Since we have begun to discuss the shining of the sun onto Earth, a number of brief declarations need to be made concerning the relation of light and shadow.

Assertions.

1. Elemental earth is not itself luminous.
2. The light of the sun penetrates into things composed of elemental earth. This is obvious by experience, for light shines through horn, wood, food, and the like when it is cut into thin sheets.
3. Bodies that are by nature opaque become more reflective with polishing.
4. Bodies that are by nature diaphanous become more transparent, and less visible, with rubbing.
5. Bodies that are by nature midway between opaque and transparent become with rubbing more reflective and more translucent.
6. Colored things tinge the light that reflects from them or passes through them. Hence a red body projects red color onto a white wall. Hence the eye gazing into a colored mirror, or afflicted by jaundice, perceives all things tinged by color.

Thus Kepler correctly reasons that the planets are translucent, and that this is why they appear to differ in color. Indeed, he makes the credible statement that the planets are all bathed by the light of the sun, but this light is affected by colors innate to those planets by nature (Francois d'Aquilon teaches this also). The light, having been so affected, then rebounds into our eyes. But if Kepler ⟨(book 1, proposition 35)⟩ grants this in the case of the planets, he denies it (wrongly) in the case of the moon.[106]

7. Whites are the most visible of all the colors. Water and watery areas are more reflective than earth and earthy areas. Hence the water portion of the globe is brighter than the land.

8. It is probable that the light of Earth reflects all the way to the moon and beyond. Nevertheless, we completely deny altogether that the ashen or secondary light of the moon is generated from this reflection of terrestrial brilliance. You will learn our reasoning in disquisition 27.

9. The portion of Earth that is illuminated by the sun shrinks as the sun recedes and enlarges as the sun approaches, but is always larger than a hemisphere.[107] Were the sun to recede eternally, the illuminated portion of Earth would continually decrease, but never reach the point at which it would be less than or equal to a terrestrial hemisphere.[108] And lest this idea be thought to proceed from an impossible hypothesis, consider the following:

Lemma.

It is possible for the sun to be moved continuously and eternally over a visible semicircle and yet not to pass through a quarter of a circle during that movement.

[See Figure 17-1.] Circle BC is drawn around center A. The sun is BD. The center of the sun is B. The sun moves from B toward C such that its center B follows the circle BC. I say it is possible *for the sun to be moved eternally and continuously in such a way that its center B may always approach C and yet never entirely reach C.*

Consider an indefinitely long straight line AE drawn from A through B into D, bisecting the sun. From point D a second indefinitely long straight line DF descends, parallel to the straight AC. Let the straight line AE be moved along DF, toward and beyond F, such that it pivots about A. Let the intersection of AE and DF be G. Let the center of the sun B be carried with AE. Since DF is parallel to AC, AE will intersect AC at A and will never be coincident with AC. Therefore point B will never reach C, for B is in line AE, and line AE has only one point in common with AC, that being A. And because the intersection G of AE and DF is continually in motion from one point to another on DF, point B is continually in motion from one point to another on arc BC, moving toward C.

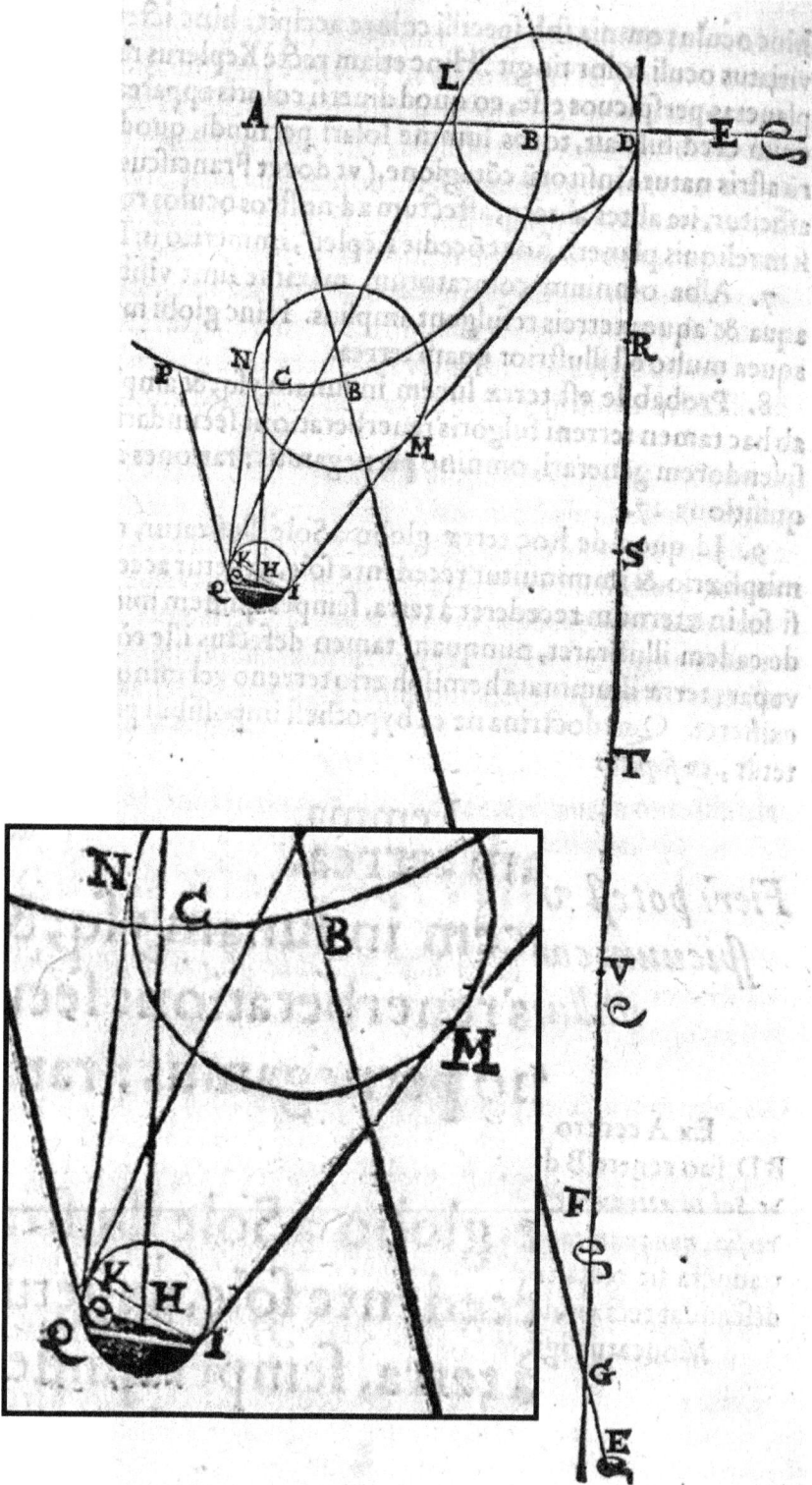

FIGURE 17-1

Therefore the sun will move eternally but will not pass through a quarter of a circle. *From which it is obvious:*

1. When the sun rose on the newly built world, it is possible that it could have ascended eternally, never reaching the meridian (noon), or it could have descended eternally, never setting.
2. It is possible that the sun could always approach Earth, yet never reach it; or always recede from Earth, yet never disappear from view or become so distant as to have no effect on Earth.
3. So, imagine the globe of Earth HIK drawn around point H of line AC [in Figure 17-1]. The sun illuminates less of Earth by means of rays LK and DI than by means of rays NO and MI. As the sun approaches nearer and nearer to point C (into infinity) it will illuminate more and more of Earth at H.
4. Let arc CP be equal to the length of straight line BD. Then extend from P a straight line PQ, tangent to Earth at Q. The farthest ray of the sun NO will reach neither beyond P nor below Q, nor will any other ray, for the reason that B will never arrive into C.
5. Hence the infinite divisibility of quantity is established to be a thing not only of the mind or of the imagination, as certain mathematicians think, but of nature itself. Indeed, it is established in physical things. For (into infinity) the arc of Earth KQ is illuminated toward Q by the tangent ray, and the length of illuminated arc is somewhat increased. Thus, by each moment the illuminated hemisphere of Earth is enlarged by the approach of the sun and the dark hemisphere is diminished.
6. It is no less obvious that *Nature* has recourse to the smallest physical limits. Indeed as AE, carried upon straight line DF, passes through equal intervals DR, RS, ST, TV, etc., smaller and smaller intervals answer back on arc BC, into eternity, and smaller and smaller portions on arc KQ are illuminated, increasing the illumination of the whole Earth.

DISQUISITION 18.

Concerning air and fire.

The air, the place of weather, encircles the terrestrial globe. For here the exhalations of Earth and the evaporations of water accumulate into a sphere that, according to its disposition, tempers variously the light arriving from celestial bodies. Indeed, this air brings forth from the rays of the sun the dawn and the dusk. It stains the heavens scarlet, both when the dawn breaks and when the northern lights shine. It scatters the shooting stars and the heats. It shows the sun and moon sometimes pale, sometimes ruddy. It generates halos and the like. Indeed, these things arise from the infinitely varying colors, densities, and depths of those vapors, which are continually agitated and almost never at rest.

Thus, whatever some may say to the contrary, these vapors are the sole cause of that trembling splendor seen in all stars called scintillation or twinkling, inasmuch as the churning of the vapors between us and the stars disturbs the rays of the stars. Reason, applied to the evidence arising from innumerable common experiences, demonstrates this to be so. It is on account of this (all else being equal) that the same star twinkles more strongly when near the horizon and more weakly when near the zenith; that the fixed stars twinkle more vigorously, the wandering stars more feebly; that stars twinkle more frequently under a northern wind, less under a southern wind. All arise from the restless vapors.

These same vapors cause the sun and moon to be oval in shape when rising and setting, by means of refraction along a line perpendicular to the horizon, brought about by the fact that these vapors have a spherical shape concentric with Earth. From which it also follows:

1. That the sun and the stars appear higher above the horizon than they truly are.
2. That they appear to rise earlier and set later than they truly do.
3. That they appear closer to each other in altitude, distorting their relative positions.
4. That which is above the vaporous or atmospheric region—above the air we breathe—is far more pure and refined. This some call aether, others fire, although it seems to have little in common with our fire. Indeed, for many reasons they correctly deny our fire to exist above the atmospheric region.

And indeed this is all concerning the parts of the sublunar world, beyond which lies the heavens. All these compose one ornate, elegant, pleasingly ordered and visible whole called *The Universe*.

DISQUISITION 19.

Concerning the heavens in general.

The stars give evidence that a substance and machinery of the heavens exists. First, the stars are seen, and their light and power could by no means reach us through a vacuum.[109] Second, a plurality of motions is observed in the stars.

Philosophers of old wanted both an innumerable multitude of stars (and each a separate world) and an immeasurable magnitude of heaven. But we thoroughly deny the existence of any infinity, either in multitude or in magnitude. In disquisition 10 we showed such infinity to be not possible. Now we further demonstrate this.

Let there be infinite contiguous straight equal feet AB, BC, CD, DE, EF, etc., beginning with A and extending to the right, as seen in the top row [of Figure 19-1]. Half of each foot is white (GH, BI, CK, DL, EM, etc.). Half of each is black (HB, IC, KD, LE, MF, etc.).

Suppose God eliminates the black halves, leaving the white halves remaining, as seen in the second row [of Figure 19-1]. Next God joins together the whites (third row). Alternatively, God might repeat one of the whites GH as many times as there are white segments in the universe. Or, as another alternative, God might stretch out some infinite line NS, etc. (bottom row) in which each part NO, OP, PQ, etc. corresponds in size and number to the white halves AH, BI, CK etc. that were left behind.

The placement of the conjoined halves AH, BI, etc. (or of the line NS, etc.) will be either toward G (as in the third row), or toward F (as in the fourth row). If it is toward G, then beyond all the conjoined white half-feet AH, HC, CK, KE, EM, etc., will remain in the space MF the length of the removed black half-feet HB, IC, KD, LE, etc. (equal to the space of the white halves).

Either the length of the conjoined white halves is infinite, or it is not. If not infinite, then the whites are not infinite in number, which is contrary to the initial statement. If infinite, then suppose God were to bring back the demolished black halves and place them in the space MF owed to them. Then an entire infinity AM would be contained within the closed limits G and M, which is impossible.

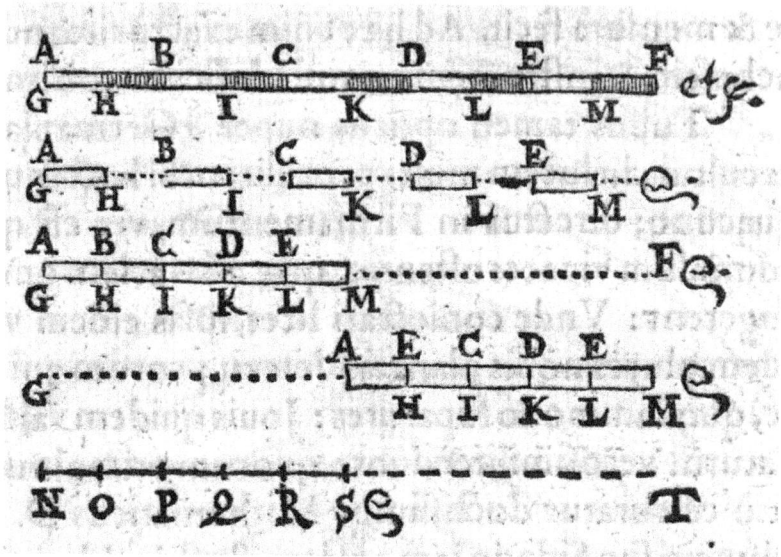

FIGURE 19-1

The same will happen in the case of line NS, etc. It will not be equal in length to the original whole AF. The length of feet formed by the remaining white halves will not equal the whole.[110]

If the placement of the conjoined white halves is toward F, as in the fourth row, then every length of every annihilated black half is fit between point G and half-foot AH, and again an infinite distance GA is enclosed by boundaries G and A, which is impossible.

Therefore an infinite extension is impossible to be granted. Therefore neither an immeasurable heaven, nor innumerable stars, will exist.

DISQUISITION 20.

All the same, only he who has created all things—according to number, weight, and measure—may number the multitude of the stars and measure the vastness of the heavens. Indeed, for the purpose of examining these things exactly, human sight grows feeble and all instruments fail before such immenseness.

Nevertheless, the optic tube, which has come to Germany, improved by Italy, has brought forward into the light many things unknown until now. Direct it toward any point in the firmament, and it may reveal some lively little stars that cannot be seen by even the keenest eye—stars which, we might conjecture, are very far away.[111]

This tube has revealed several planets, understood to be attendants or satellites of Jupiter and Saturn: four noble attendants for Jupiter, two for Saturn. ‹Galileo is their discoverer.› That most learned mathematician Galileo Galilei, now distinguished for some time by his *Starry Messenger*, is rightly celebrated for the first glory of discovery of these. This same tube has shown Venus to emulate the moon, has shown the moon to be uneven, and has shown the sun to be spotted. And from day to day it reveals to Earth other delightful, useful, and wonderful things of heaven. ‹Such is the fruit of the optic tube.›

DISQUISITION 21.

There is no agreement among authorities as regards the number, order, and placement of the spheres of the heavens. Some authorities propose greater numbers of spheres, some propose lesser. In 1574, some years after Nicolas Copernicus, the noble physician Girolamo Fracastoro published ‹his opinion favoring› a homocentric[112] system of spheres—that is, a system consisting of spheres that all have one common center. His contention was that this system accounted for all heavenly phenomena by means of only spheres concentric to Earth, with all eccentrics and epicycles[113] eliminated. He attributed the increase and decrease in the apparent magnitude (that is, size) of the wandering stars (that is, the planets), the sun, and the moon not to any approach or retreat of them, nor to the air or fire that lies between us and the moon, but to celestial orbs interposed between us and these bodies that alter the light emitted from them. The varying density or rarity of an interposing orb variously reduces or magnifies a celestial body by means of refraction, thus causing it to appear to the eye to shrink or grow.

He accounts for the various motions of the planets—their progression through the fixed stars, their stopping, their retrograde motion, and so on—by introducing other orbs with astonishing motions this way and

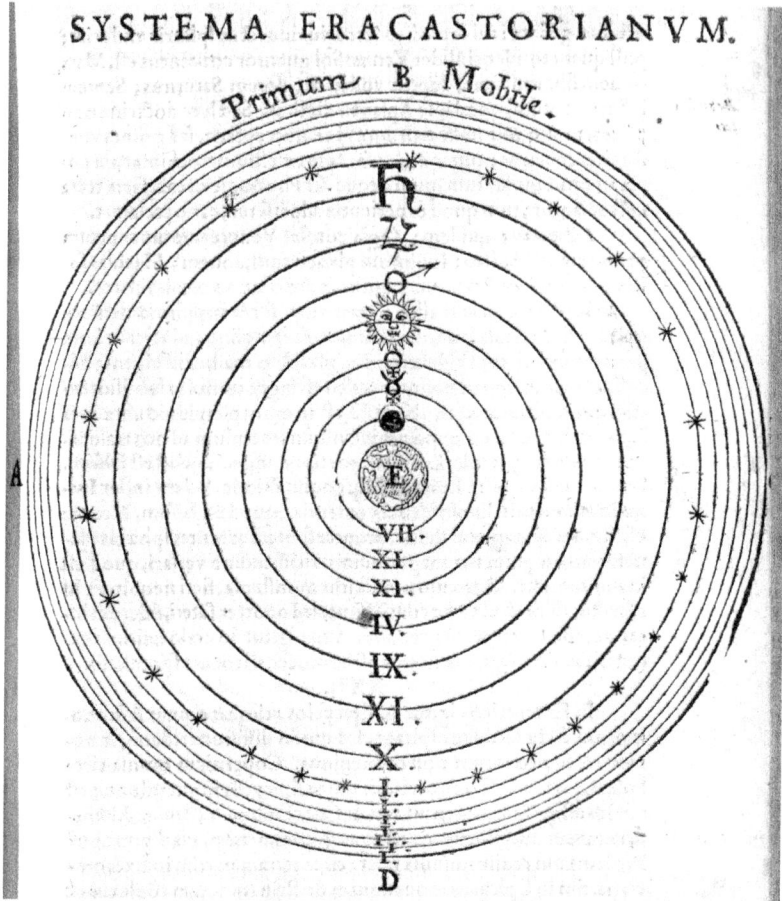

FIGURE 21-1

that, so that the number of heavenly orbs, all concentric to Earth, grows to seventy, as seen in [Figure 21-1]. Thus, in this system, the moon is carried around the center of Earth (E in the figure) by seven orbs; followed by Mercury borne by eleven orbs; then Venus, also with eleven; then the sun with four; then Mars with nine; Jupiter with eleven; Saturn, ten; and the firmament of fixed stars, seven (he still refers to it as the Unmoving).

But this doctrine does not satisfy ‹and is to be rejected,› because no astronomers accept it (they unanimously reject it), because it is burdened by such a prodigious quantity of celestial machinery, and because it fails to explain all the discovered celestial phenomena. Experience opposes it. Reason cries out against it.

Experience opposes it because the optic tube has established that the center of Venus's own motion is the sun, that the center of the motions of the Jovian satellites is Jupiter, and that the center of motion of the solar spots is again the sun. Therefore epicycles do exist in the heavens.[114]

Reason cries out against it because if the apparent magnitude or size of celestial bodies is governed through refraction by celestial orbs, then the heavens must be more refractive than our air, which is absurd. Also, the stars and planets near to the moon should be enlarged or reduced by the same proportion as the moon is enlarged or reduced, since all will shine through the same refractive orbs. Likewise, when the sun is eclipsed by the moon, their sizes should both change together.[115] All the stars across the firmament should perpetually be changing in magnitude, owing to the fact that the refractive orbs of the planets, moon, and sun are turning this way and that below them.

None of these monstrosities occur, so the changes in magnitude of celestial bodies should not be attributed to refraction alone. It should be acknowledged that these changes happen because the distances to these bodies increase and decrease, consistent with heavenly epicycles. The speculation regarding homocentrics is futile and false.

DISQUISITION 22.

All astronomers other than Fracastoro generally concur regarding the existence of celestial eccentrics and epicycles, but differ regarding the arrangement of them. Therefore astronomers do not agree on the number and placement of heavenly spheres.

Copernicus has all the planets moving around the sun by means of both eccentrics and epicycles.[116] His system has been treated in disquisition 13.

Christopher Clavius, the restorer of the discipline of mathematics who will be forever remembered, follows Ptolemy and others (such as Alphonso;[117] and the *Hypotheses of the Heavenly Orbs* by an anonymous author, published by Conrad Dasypodius;[118] and the 1489 *Compilation* by Leopold of Austria[119]) in that he has all the stars and planets moving

FIGURE 22-1

around Earth by means of both eccentrics and epicycles. In his system there exist in the universe thirty-four celestial orbs, ordered into eleven spheres (rather than ten as earlier authors said), ‹as he states on page 36 of chapter 1 of tome 3 of his *Sphere*,›[120] and as is shown in [Figure 22-1]. Earth, water, and air EF stand in the middle of the universe, surrounded by fire GH. Above these is the moon, riding on a system of five orbs. Above the moon is Mercury, on six orbs, Venus, on four, and then the sun, on three. Next are Mars, Jupiter, and Saturn, each the lord of four orbs. These are all enclosed by the firmament, which in the time of Aristotle was the eighth and highest sphere. But since then a ninth was added by the time of Ptolemy, a tenth by the time of Alphonso the King, and in our time an eleventh (on account of a motion discovered by Copernicus) by Clavius and the illustrious mathematician Antonio Magini, as is found in the same writings of Clavius.

Actually, this used to be the system of Clavius and Magini, at the time when nothing contrary to it could be found in the heavens. But in truth, after that keen dioptric tube had been turned to the heavens, each changed his opinion. Indeed, Magini is seen to construct a different celestial system — like that one found in the writings of Martianus Capella — in which the motions of the heavens are explained through some sort of retarding action. Meanwhile, Clavius said the following in his swan song (the last edition of his work, published late in his life):

> I do not want to hide from the reader that not long ago *a certain instrument was brought from Belgium.* It has the form of a long tube in the bases of which are set two glasses, or rather lenses, by which objects far away from us appear very much closer, and indeed considerably larger, than the things themselves are. This instrument shows many more stars in the firmament than can be seen in any way without it ... and when the moon is a crescent or half-full, it appears so remarkably fractured and rough that *I cannot marvel enough that there is such unevenness in the lunar body.* Consult the reliable little book by Galileo Galilei, printed at Venice in 1610 and called *Sidereal Messenger.*
>
> Far from the least important of the things seen with this instrument is that *Venus receives its light from the sun as does the moon,* so that sometimes it appears to be more like a crescent, sometimes less, according to its distance[121] from the sun. *At Rome I have observed this in the presence of others more than once.* Saturn has joined to it two smaller stars, one on the east, the other on the west. Finally, Jupiter has four roving stars, which vary their places in a remarkable way both among themselves and with respect to Jupiter — as Galileo Galilei carefully and accurately describes.
>
> *Since things are thus, astronomers ought to consider how the celestial orbs may be arranged in order to save [i.e., explain] these phenomena.*[122]

(See tome 3 in chapter 1 of his *Sphere of Sacro Bosco,* page 75.) ‹Thus Clavius abrogates his system.›[123]

That system of Clavius *does not sufficiently explain the given phenomena,* when all astronomers are certain that Venus, because it imitates the moon as regards its manner of shining, goes around the sun (as does

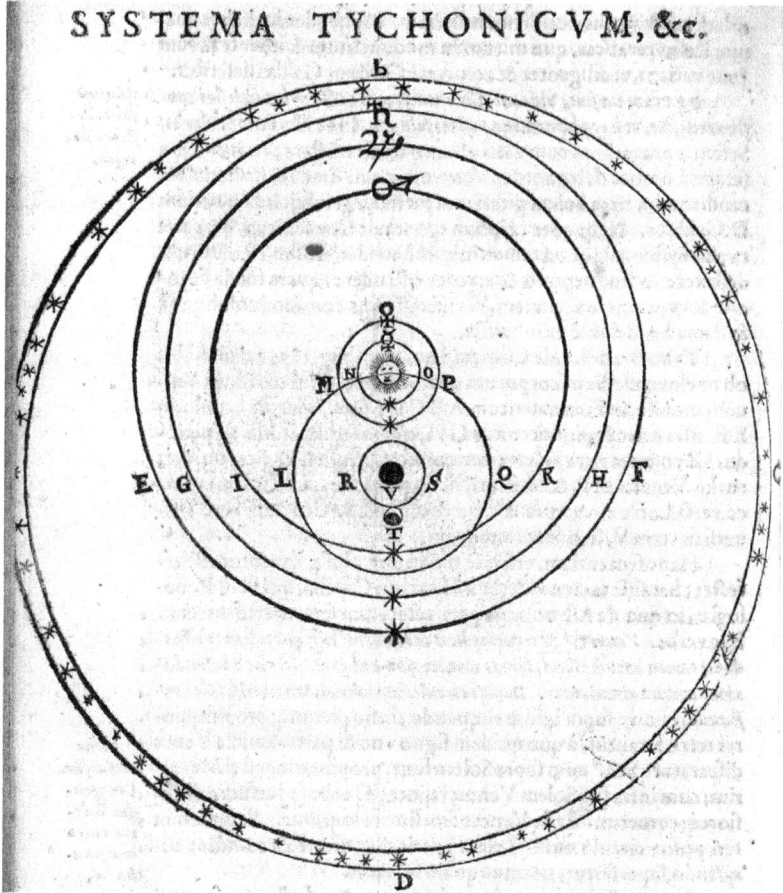

FIGURE 22-2

Mercury, according to Galileo, as I have read). And in view of that, Copernicus cannot be ignored, as is obvious from his system. Nevertheless, his system pleases very little. Without doubt it is easier to demolish and to show errors in this business than to build solidly and to reveal the truth. Had Clavius lived, no doubt he would have produced some suitable system.

Tycho Brahe, on page 189 of his 1577 second book on comets, arranges the celestial bodies as in [Figure 22-2]. The moving edge of the universe is the firmament of fixed stars ABCD. Within this is Saturn's circle of revolution around the sun EF, inside of which is Jupiter's circle around the sun GH, followed by that of Mars about the sun IK. Next comes the circle of the sun around Earth LQ, and in turn the circles of Venus MP and Mercury NO around the sun. Last is the orbit of the moon RS around Earth V. Earth is stable and unmoving, truly the center of the universe.

Although Brahe dissembles and pretends to be the author of this idea, he seems to have appropriated it from Martianus Capella. Capella, who discusses astronomy in book 8 of his *The Marriage of Philology and Mercury*, says the following:

> *Now Venus and Mercury, although they have daily risings and settings, do not travel about the earth at all; rather they encircle the sun in wider revolutions. The center of their orbits is set in the sun.* As a result they are sometimes above the sun; more often they are beneath it, in closer proximity to the earth. Venus's greatest elongation from the sun is one and a half signs. When both planets have a position above the sun, Mercury is the closer to the earth; when they are below the sun, Venus is the closer, inasmuch as it has a broader and more sweeping orbit.[124]

Regarding Venus, Capella also says, "Located on its own epicycle, it goes about the sun, varying its course. . . . at times it is borne above the sun and at times beneath it."[125] In the same book, under a heading announcing that Earth may not be the center of the planets, Capella writes, "This general observation must be made, that the earth is eccentric to the orbits of all the planets (that is, it is not located at the center of their circles)."[126] Regarding Mars he says, "Mars has its own course, beyond the sun, and revolves about the earth, which is eccentric to its orbit,"[127] and regarding Jupiter, "Its ascents and descents prove that its orbit is eccentric with respect to the earth."[128] And we may infer the same concerning Saturn. Thus the hypothesis of Brahe, in which the five planets circle the sun which in turn circles Earth, is not much different from that of Capella. This is especially since true since at the end of book 8 Capella also writes,

> The powerful effect of the sun's rays is responsible *for the anomalies in the orbits* of all the aforementioned planets and for their stations, retrogradations, and progressions. The rays strike the planets, causing them to rise aloft or to be depressed. . . .[129]

This certainly indicates that, in the opinion of Capella, the sun is the master and center of the five planets. Thus it seems most probable that Tycho constructed his system based on that of Capella, adding very little.

I do not know whether the Capellan system is that which Clavius called for. Certainly it is appealing to many astronomers at this time because it is very consistent with the heavenly phenomena, and it shares nothing with Copernicus and much with Ptolemy. It removes superfluous celestial movement and explains everything easily by means of fewer orbs.[130] Furthermore, at one time ancients shared the opinion of Capella. Indeed, Vitruvius writes ‹in book 9, chapter 4›, "But Mercury and Venus, their paths wreathing about the sun's rays as their centre, retrograde and delay their movements. . . ."[131]

DISQUISITION 23.

So the order of the celestial orbs follows the order of the wandering stars, which is, from nearest to farthest: moon, sun, Mercury, Venus, Mars, Jupiter, Saturn. This order is the same as that in the writings of Clavius, tome 3, page 42. But if we consider their centers of motion, the moon circles Earth; Venus, Mercury, and the other three wanderers all circle the sun (which the moon-like appearance of Venus has rendered certain); and fewer orbs result than if we reckon them all as circling Earth.[132] Yet the properties of these orbs (number, material, thickness, and the like[133]) are not yet established. At present nothing is settled concerning these.

Neither the apparent nor the true magnitudes or sizes of the wandering or fixed stars are clearly evident, with different observers discerning and judging those sizes differently. The cause of this variation is (because greater and lesser accuracy of observation in estimating stars is of importance to apparent magnitude) difference of vision. For the stars generally appear smaller to those who have very good eyesight, and larger to those whose eyesight grows feeble. Differences in judgment regarding apparent size transfer over to differences regarding true size. Thus, differences of opinion arise from differences of sight. ‹This is why writers disagree regarding the magnitudes of stars.›

And note the matter of the different colors of the stars, which works in a strange manner, enhancing this or that distant star and causing error.

It is a notable phenomenon that those stars that are whiter in color impress themselves upon the eyes in such way as to appear larger than those stars that are of a redder hue. The optical tube instructs us of this illusion. It faultlessly reveals the bodies of the stars, showing the star Arcturus to be no smaller than Sirius the Dog Star. ‹This is a remarkable phenomenon.› Who would have believed it to be true? The reason for this thing is because *all white things strike the eye strongly and stun the sense of vision, and so are seen as if they were alone, with everything else extinguished.* This can be tested with a glowing coal and a candle flame. The candle flame may be smaller than the coal, but nevertheless, seen from a distance it will appear larger. There are many other examples of this.[134]

DISQUISITION 24.

The diversity of colors in the fixed and wandering stars comes not from the sun but from their own internal composition. They themselves emit no light and are only illuminated by the sun. ‹See chapter 30 of Al-Batani and book 4, paragraph 77 of Vitellius.›[135] Thus they all receive the same kind of light. Nevertheless they temper the light in which they are bathed, each star according to its own physical constitution, and thus they send back to us here below one or another color in varying strengths. For the heat passing out of an iron oven feels different in our bodies than the heat passing out of a brick oven. Likewise, the rays of the sun are tinged differently by ice than by water or thin wood—and the difference is not ascribed to the sun or to its light, but to the body onto which the light from the sun has fallen.

The optimal method (used regularly now for three years) of truly discerning the genuine colors of the ‹fixed and› wandering stars[136] is to remove the concave or eye glass or lens from the optic tube, so that only the convex or object glass or lens remains. Then the natural color of a star pours through best, by direct sight on the star. But the star appears mangled and not at all whole, this being attributed to the passage through the narrow tube—the more tube the light passes through, the more this happens.[137]

There may be dark patches or bright gleams that can be attributed to bubbles in the glass. There are several signs of this. One is that all eyes see the same thing when using a certain lens, even at different times. Another is that different lenses show different things in the same star. Yet another is that if the lens is rotated, these dark patches and bright gleams are carried around with the rotation, not changing position among themselves. If the gleams shine constantly and maintain their positions, as happens frequently, then they can be attributed to the lens. If they boil up and vanish at intervals, if they move around, if they grow and shrink in size and clarity, if they alter shape—as happens not rarely—then these gleams can be attributed to the eye, and in particular to drops of moisture running across the eye, especially if winking or rubbing the eye promptly scatters them. The twinkling or scintillation that is observed in a star is also important to keep in mind.

Concerning the Types of Stars
(or Heavenly Bodies)[138]

First, concerning the moon.

The moon is the closest of all the planets that revolve around Earth. It always directs the same face toward Earth, as is established by the spots seen on its face. In the opinion of all, this occurs by means of an eccentric, an epicycle, and a monthly rotation of the body of the moon about an axis through its poles. Although these motions are circular, the result is a libration or rocking of the moon about a line extended perpendicularly through the center of the moon to the center of Earth.[139] These all are shown as follows [see Figure 25-1]:

Let the moon's deferent circle BCD be described around center A. Around point C on the periphery of the deferent let be described the epicycle circle EFGH. Let be described at point E on the circumference of the epicycle, the new moon; at F and H, the quarter moons; and at the apogee point G, the full moon.

Because the center of the moon is always on the circle EFGH, by necessity some portion of the moon will always fall inside that circle, the other portion outside that same circle. If rays are drawn tangent to the moon from an eye on Earth at A—namely AI and AK, AL and AM, AN and AO, AP and AQ—they will show what parts of the moon can be seen

63

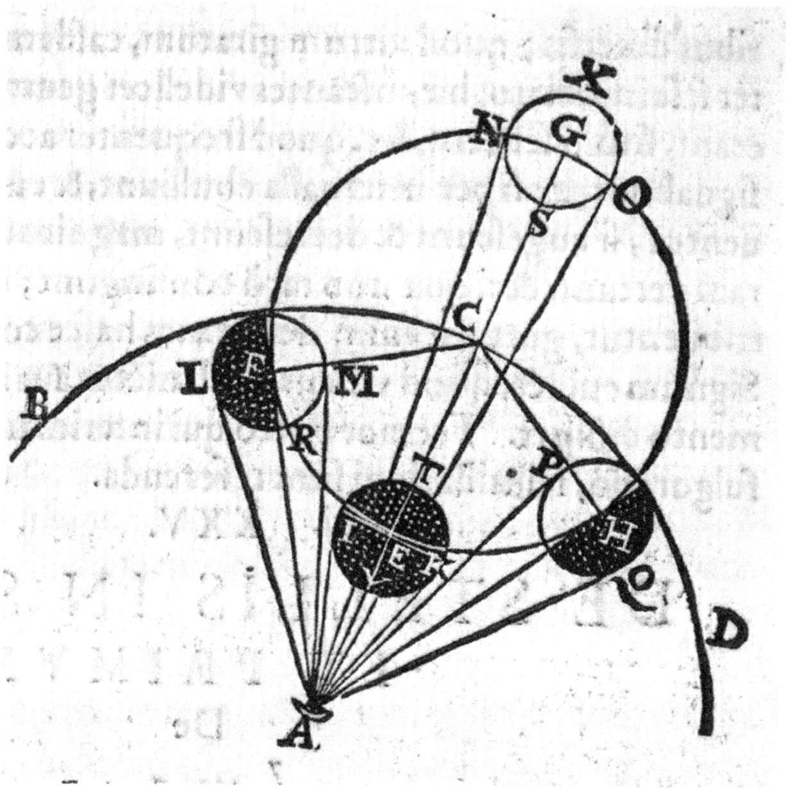

FIGURE 25-1

inside and outside of the concavity of the epicycle. Indeed, were the straight line IK, the visible diameter of the moon, regarded as tangent to the epicycle at E, it would fall entirely outside the epicycle, as would the whole hemisphere IVKE. Therefore, that which the eye sees between the tangent rays AI and AK will be all the more outside the epicycle, since that will be smaller than the IVKE hemisphere, on account of the closeness of the moon at E to the eye at A.

Since ray AF is tangent to the epicycle at point F, which is the center of the moon, the periphery of the moon will cut the epicycle radius FC at M. Line AM is therefore inside the epicycle, tangent to the part of the moon that lies within the epicycle. Therefore, the part of the moon that is visible from Earth and that is inside the epicycle is MR; the part that is outside is RL. Therefore, the entire visible portion of the moon NSO is inside the epicycle, and so forth.

If the moon only circles the center of its epicycle C, itself unmoved (like a wart affixed to that epicycle), and does not rotate in an opposite direction about its own center (E, F, G, or H), then any part of the moon that is outside the epicycle at one point will always be outside. Thus, the part which is NXO will be the same as that which is IVK, and vice versa, while NSO will be the same as ITK, and so on. The IVK seen at E is the NXO that is unseen at G; the MRF observed at F is hidden at H, and vice versa. Therefore, if the moon does not turn about its own center, it will always show a different face. But daily experience contradicts this, since the visible part of the moon always appears the same. The spots visible on the moon, clearly seen by all both now and in ancient times, testify to this.[140] Therefore, the body of the moon is not fixed but is mobile. Indeed, Kepler himself has said ‹in *Optics*, chapter 6›,

> in going around the earth, [the moon] presents one and another part to the sun, as if it were to be turned by itself as if on a spit, roasting itself every month towards the sun, in very nearly that manner in which Copernicus said that the earth is turned and roasted daily towards the sun, as towards a fire.[141]

DISQUISITION 26.

The primary light of the moon.

The light of the moon is of two kinds: a primary and a secondary. The moon shows the secondary or ashen light from new moon to eight days past new, in that part of the moon that does not receive the rays of the sun, and it shows the primary light in that part that is exposed to the sun. Truly the moon itself produces no light, as is seen when it is fully eclipsed and goes utterly dark, as we and others have observed. Therefore all lunar light, whether from a crescent moon, or a full, or a half-full, or a gibbous, must necessarily be from an outside source—the sun. ‹See Thales of Miletus;

see Anaxagoras of Clazomenae; see Vitruvius, book 9, chapter 2; see *On Scipio's Dream* by Macrobius, book 1, chapter 19›.[142]

The moon receives this light not in the manner of a mirror, but in the manner of a cloud, taking it in and throwing back part—and throwing it back not as from a polished, geometrically perfect spherical body (otherwise it would not be possible for the whole moon to appear bright[143]), but as from an astronomically spherical body.[144] That the moon is indeed a sphere is proven from the border between light and shadow, which sometimes is a convex arc, sometimes a concave one (and this arc is always the shape of half the perimeter of an ellipse), and sometimes a straight line.

Some observers notice that the illuminated edge of the moon will stand out beyond an adjacent dark edge, as though rays from the illuminated edge were being thrown out farther. This is what the feeble eye sees. But experience shows otherwise to the keen eye, and to the optic tube, and to the learned. For a smaller white post standing next to a larger red one appears larger when seen from a distance, yet it is not. Likewise, on a striped sheet of paper, half black, half white, the white stripes appear larger, the black smaller. *A filament of spiderweb illuminated by the sun gives the appearance of a silver thread ten times larger than a filament that is not illuminated.*

Consequently, a perverse reasoning arises that says that the shining moon grows because of the darkness. This is to be rejected. Indeed, *the cause of this is not the darkness, but the eye,* which sees brighter things as larger. *As shown above, whites and brights overpower everything else* and thus conquer the eye.

But all these things will be better understood from the detailed drawing of the moon provided here [Figure 26-1], which includes the following: *First,* lunar spots known since antiquity, marvelous in variety (A, B, C, D, E, F, and G). *Second,* newly discovered spots[145] (H, I, K, L, and M) whose forms change over time, proceeding from crescent shaped to round, like so many little moonlets, by reason of shadows cast by the higher parts. *Third,* the distinctly uneven lunar light—in the spots known since antiquity, where gems of intense whiteness such as N and O are manifest; likewise in the spots that are like moonlets, at H, I, and K, and near L, M, and D and innumerable other gems. *Fourth,* unchanging black spots, at P, Q, R, and elsewhere. *Fifth,* the boundary of light and shadow, STVX, always convo-

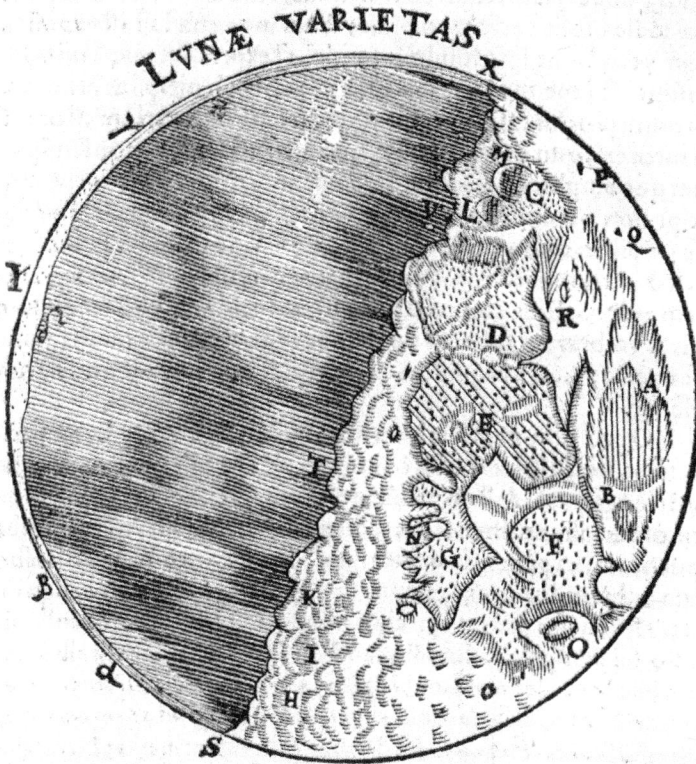

LVNÆ VARIETAS

FIGURE 26-1

luted, rough, and uneven because the higher parts unevenly block the light of the sun. *Sixth*, the shadowy part of the moon XYS, the whole thing shining by secondary or ashen light (see the next disquisition) whose luster is comparable to Saturn. *Seventh*, the three levels of light that are found in the shadowy part: the ordinary light, generally equal everywhere, present in the expanse STVXZδ; a more intense light, in the ring designated XZYαS; and surpassing that, a light in the form of a sprinkling of luminous points, in that part of the ring designated βY$\gamma\delta\beta$.[146] These points have been carefully observed many times by several people through average optic tubes (which in this case agree with the most excellent tubes) on 1 October, 27 November, and 29 December 1612, and also 23 April and at several other times in 1613. They are most likely to be seen around new moon. They present an appearance similar to the tiniest stars in the firmament. Skillful diligence of observation is required to see them.

DISQUISITION 27.

The secondary light of the moon.[147]

Many are of the opinion that the secondary or ashen light of the moon is rather bright — that such light can be detected even before the sun is fully set, something that does not happen even with all of the first-magnitude stars. I say that if the primary light of the moon is hidden behind the roof of some house, it is not unusual for the secondary light to shine forth with the luster of a star of the third or second magnitude — certainly the equal of Saturn, if not surpassing it. This I have confirmed through three years of frequent experience. Thus a most learned mathematician, encountering this glow by chance, might wonder whether this might be the full moon at the wrong time.[148] Because of this I say the following:

1. *The secondary light is not innate to the moon itself.* The common opinion of most mathematicians is that it is not. Otherwise, it would shine forth during an eclipse of the moon, and that does not happen. Indeed it is evident that the moon, when covered by the middle of the shadow of Earth, is thoroughly devoid of light and color. The rusty color presented by the moon when it is passing through the edges of the terrestrial shadow is nothing other than the light of the mostly-eclipsed sun, which passes around the edge of Earth, falls onto the moon, and is reflected back to us, as is demonstrated in optics.

2. *The secondary light is not the result of Venus or another star illuminating the moon.* The majority think it is not because the secondary light shines forth all the same in a conjunction of Venus and the sun. For if Venus is on the near side of the sun, between the sun and Earth, all the radiance of it is directed away from the moon and toward the sun.[149] If Venus is on the far side of the sun, so that the sun is between it and Earth, then the radiance of Venus falls on the face of the moon that is turned away from Earth, not on that turned toward Earth. For

like reason the more distant stars are even less able to be the source of the secondary light.[150]

3. *The secondary light is not the result of Earth illuminating the moon.* See Macrobius's *On Scipio's Dream,* book 1, chapter 19 and elsewhere, against certain recent ideas on this.[151] Indeed, what could bestow to Earth such power that to us it illuminates the moon in the amount that the sun illuminates Saturn? Saturn must without doubt outshine Earth, and the secondary light of the moon compares to the light of Saturn and is not less than that light. "Above the aether," Cardano says ‹in book 3 of his *On Subtlety,* near the beginning, regarding the moon›, "all things shine and glitter to such a degree that, if indeed we might inspect the moon, being there during the time of an eclipse, we might be blind on account of a splendor not unlike innumerable brilliantly lit candles focused into the eyes."[152] But the secondary light of the moon greatly surpasses the rusty glow seen in an eclipse. What of that then? And this light is from Earth?

In addition, who would deny that Venus has as much power to illuminate Earth as Earth to illuminate the moon? And yet does not Venus beam very little radiance toward Earth? In fact, all the stars together (in the absence of the moon and sun) in no way confer as much light to Earth as Earth would confer to the moon, were Earth the source of the secondary light. To be sure, to a man placed on the moon, Earth would be an adequate light for reading. But who here on Earth can read by the light of the stars? Thus, on the moon you could read by the light of Earth, but on Earth you cannot read by the light of the stars, and so the splendor of Earth would surpass the light of the stars.

And, following this idea, it might have to be conceded that the splendor of Earth would likewise surpass the splendor of the moon. And thus the mind and the senses would have to admit that *the secondary light of the moon, received from Earth, is more intense than the secondary light of Earth, received from the moon*—or, simply, Earth pours forth more splendor than the moon. I think no one will grant such a paradox.[153]

As the moon waxes from new, its ample secondary light decreases in size but persists otherwise unchanged, all the way through to the ninth day past new. At that time it is seen, or distinguished from the primary light, with more difficulty. This is owing to the primary light itself, which overcomes the eye with its power and obliterates the rays of secondary light. As the primary light of the moon progresses from new toward full as seen from here on Earth, the primary light of Earth would progress from full toward new as seen from the moon. And as the moon proceeds from new toward full, it illuminates the nocturnal earth with increasing intensity; as it proceeds from full toward new, this illumination diminishes. Therefore Earth should do the same for the moon. However, experience shows this not to happen: we see the secondary light of the moon to diminish in space, not in luster. Therefore the secondary light of the moon is not of Earth.

Furthermore, light from Earth would strike the middle of the moon more perpendicularly and therefore would illuminate the moon more strongly, while it would strike the extremities more obliquely and therefore would illuminate more faintly.[154] But in truth, the opposite happens in the moon—witness the ring around the lunar edge. Thus it is impossible for the secondary light to be of Earth. I need say nothing about the sprinkling of luminous points.

Would anyone believe that a man on the moon would see part of Earth illuminated by the moon?[155]

Finally, the secondary light of the moon is manifestly visible in a solar eclipse along that part of the moon which eats into the sun, while the rest of the moon is entirely empty and remains black. ‹See Vitellius, book 4, paragraph 77; see Reinhold's addition to Peurbach's work.›[156] Therefore, that light is not from Earth. Experience testifies that, if it were, very little would be seen in the part that advances into the sun, and most would be seen in the other part, because a lesser light positioned directly against a greater light is not discerned, but it may be seen when nearby. Therefore, the secondary light of the moon does not originate from Earth.

4. *Therefore, the secondary light of the moon is produced by the sun,* not so much by the light of the sun shining upon the moon and reflecting from it (in the manner that produces the moon's primary light), but by the light of the sun shining into it and through it. Thus the sunlight is imbibed by the moon and then reemitted, so that the whole moon is made a quasi-lucid body, much like a cloud or a crystal.[157]

Such has been the opinion of philosophers both ancient and new. See Caesar Germanicus in the *Phenomena* of Aratus, toward the end; see Cleomedes; see Vitellius's *Principles of Optics* ‹book 4, paragraph 77›; see the *Optics* of Fr. Francois d'Aquilon ‹book 5, proposition 56›;[158] and so forth. "The moon," says Macrobius in *On Scipio's Dream*, book 1, chapter 19, "though of a denser substance than the other celestial bodies, it is still much purer than the earth, and it permits the light to penetrate to such a degree that it sends it forth again."[159]

This opinion is supported primarily by experience. When the moon is a very young waxing crescent or a very old waning crescent, the non-illuminated part shines forth much, especially at the edge. Likewise, when the moon is eclipsing the sun, the edge shines forth, as Vitellius and Reinhold testify. Indeed, many here in Ingolstadt perceived this most brilliantly in the recent eclipse of the sun on 29 May of the year 1612, when the lunar ring shone forth mightily.[160]

This opinion is also supported by reason. The secondary light is neither from any star, nor from the moon itself, nor from Earth. Therefore it is from the sun. Those things which are observed to happen are consistent with, and their causes can be reasonably attributed to, this. And indeed, as is shown in [Figure 27-1], the solar rays AB and CD drawn through and near the lunar center are longer than ray GH, drawn through the lunar edge. Hence the former are weakened more, while the latter passes through with more strength retained. Thus these edge rays form that brighter ring on the lunar extremity [in Figure 26-1], while the other rays account for the weak secondary light in general. The reason for the little points of light [in Figure 26-1] that shine forth is that there are parts within the moon more permeable to the sun's light (that these are not from the glass of the optic tube is determined by moving the tube across to other parts of the moon and observing that the points do not follow). In an eclipse of the sun, a ring is observed at the edge of the moon in that part of the moon that passes in front of the sun, and not observed in that part that is outside the sun; the reason is that in the former the solar rays are directed toward our eyes, and in the latter they are not.

From all these are established a number of conclusions.

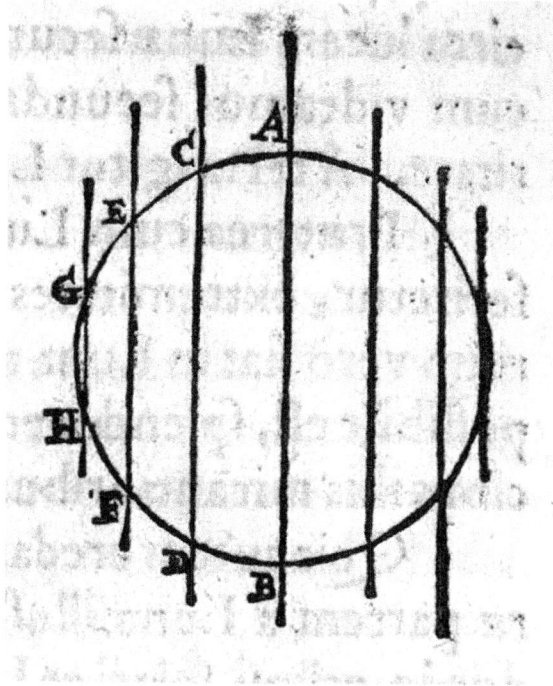

FIGURE 27-1

Conclusions.

1. Earth is not a star by reason of light, since it may shine much too feebly.
2. The moon is not an Earth, nor does it have earthy and watery parts.
3. The moon is certainly not habitable nor the abode of life. It is the mother of nothing that grows.
4. The moon is neither completely diaphanous, nor entirely opaque, but some of each.
5. At those times when the secondary light of the moon appears either more or less diminished, all else equal, this should be attributed to factors other than the moon itself.
6. It is a fiction that the bright ring around the moon is caused by air.
7. The darker parts of the moon are not watery. If they were, then they might be more translucent or more apt to reflect light.

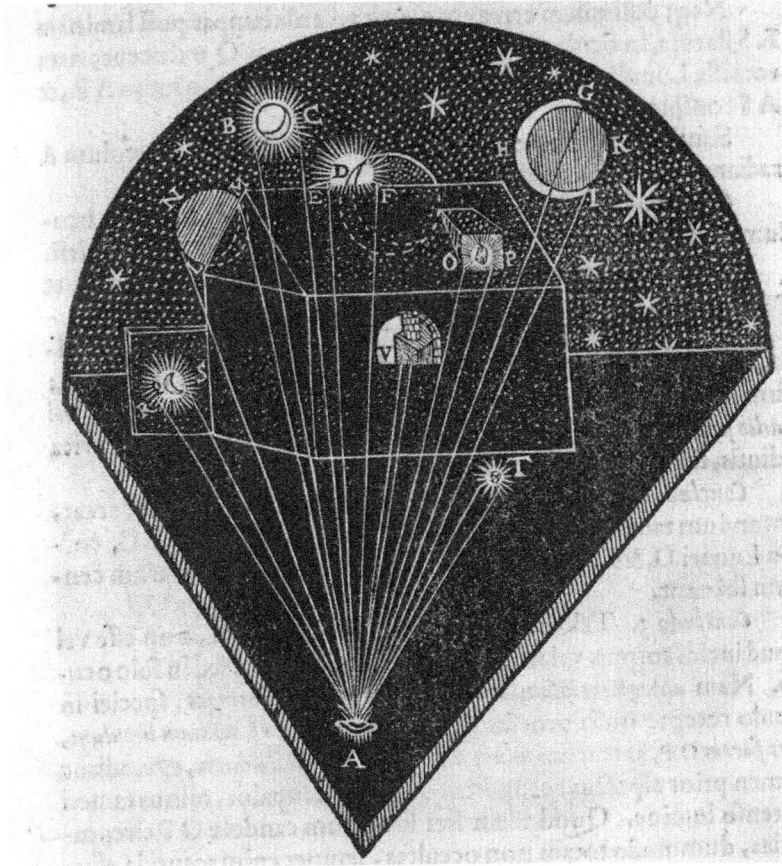

FIGURE 28-1

DISQUISITION 28.

Experiences of light.

So that all these things (along with many others discussed and still to be discussed) may be understood better, behold [Figure 28-1]. In this figure is seen an eye at A that, receiving rays AB and AC, beholds Venus BC to be round, but nevertheless Venus is the image of a crescent moon (here it being between the sun and Earth), as the optic tube reveals. That same eye, seeing the glow (indicated by rays AE and AF) of the tip D of the crescent moon ascending behind a house, mistakes it to be round and mistakes it for one of the planets. That eye decides the sickle GHI of a young crescent moon to be truly projecting from the shadowy region GKI. Likewise, the eye regards the dark part of the waning moon LMN, seen by itself (the bright part LML hidden behind the roof LFT), to be more radiant than the brighter stars. The eye reckons as round the oblong flame OP of a burning candle, seen from a distance through rays AO and AP, and it begets another, similar error: a lit lamp set behind a translucent screen RS, shining through a crescent-shaped hole Q into that eye at A, will be formed into a lucid orb seen through the rays AR and AS. A similar appearance arises from a red-hot coal T radiating light into eye A—indeed the coal takes on a round appearance. In the same way, the glow emerging from a little crack V in a furnace is seen as full and round in shape.

By testing from these and innumerable other obvious experiences here and there, I conclude:

1. It is a perilous line of argument to say that because the luminous bodies of the stars appear to have one shape or another, therefore that is their true shape. *For unless they are seen close at hand and free of rays*—that is, by the most excellent of eyes or by the aid of the dioptric tube—no conclusion can be reached about their true shape.

2. It is possible that some stars may appear round, yet may not be round. Crescent-shaped Venus BC points this out, as does the lunar tip D, and so on. The solar spots—if they may be counted as wandering stars (that is, planets closely circling the sun)—are not round.

3. Such conglobations of radiance originate neither at the luminous body nor in the diaphanous medium between the body and the eye, but in the eye alone. For *if with your hand you block the middle of the little moon-shaped hole Q, or of the flame OP, or of the incandescent coal T,* still the eye at A beholds a round shape, just less intense in light than before you obstructed these things. The same will happen should you encircle the candle OP with a hand: provided that you do not hide the whole, it will always shine round. From this it is clear that the halos, rainbows, and the like that appear around a lit candle

are attributable usually to the eye, and rarely or truly never to the surrounding air.

DISQUISITION 29.

Concerning the sun.

The sun is the principal of all the heavenly bodies. He is of advanced importance. He is a king, manifesting his majesty sufficiently by his light and his size, standing at the center of the five planets.[161] Indeed, they are like compliant slaves, running about this way and that way along the path of the ecliptic, while the sun himself always advances on that path step by step at a steady pace. He lacks for nothing. The attention of the others is directed wholly toward him. He reaches out to all, distributing generously from his abundance. Indeed, he himself illuminates all things by his fire. Without him the stars themselves would lie in darkness. No light which is not of the sun descends from the heavens into this world here below. His size is twice that of the new star seen in the constellation Cassiopeia in 1572.[162] He is a sphere whose volume exceeds that of Earth by at least 140 times, according to the best judgments of astronomers.

As the sun is first among the heavenly bodies, it is also first in abundance of phenomena (discussed in recent years first by Apelles and then by Galileo[163]), the daily attention to which will inform anyone of the liveliness of the sun. Please refer now to [Figure 29-1] as I briefly describe these.

An eye located at A on Earth sees daily wonders when observing the sun, especially when aided by the keen vision of the optic tube.[164] It beholds the sun to be elliptical in shape when it is rising B and setting C. And regularly during rising, occasionally during setting, the eye finds that the sun, when on the horizon, will be mangled and have jagged edges and at the same time be trembling energetically. The eye will always find the sun to be sprinkled everywhere as if by dark spots and bright blazes.[165] The spots are represented on the noontime sun in [Figure 29-1] by f, g, h, and i, while the blazes are represented by a, b, c, d, and e. Since the year 1612 many people have seen innumerable things like these on the sun.

These wonders will be explored in the disquisitions that follow.

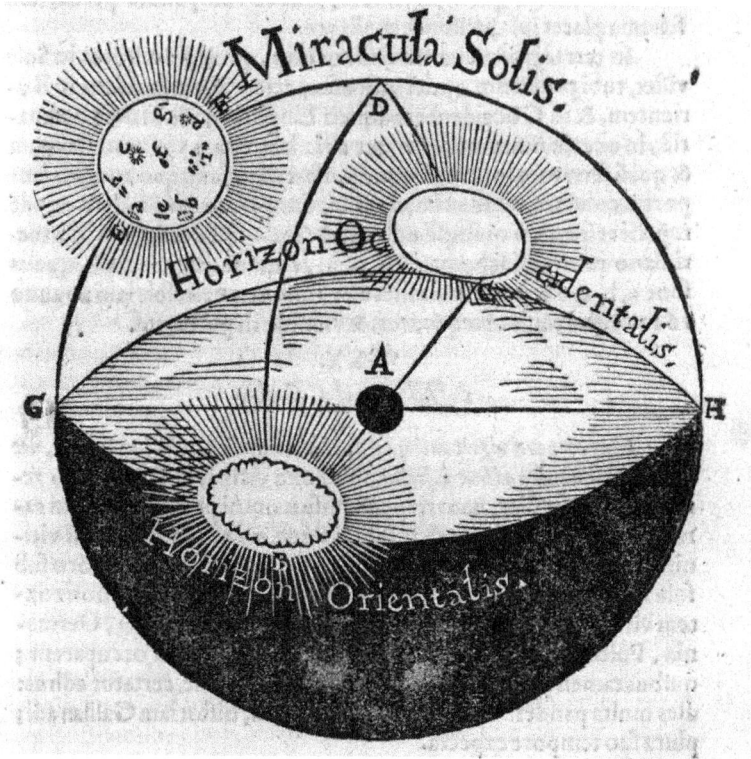

FIGURE 29-1

DISQUISITION 30.

The spots on the sun.

The spots are blackish bodies, roving around the sun by various motions. Thus far neither their number nor their nature has been determined. They are so close to the sun that they seem to be attached to it. They are utterly devoid of any parallax against the underlying sun, and thus it is evident that they are very close to the sun and not floating in the air. If they were not close to the sun, then they would not remain in the same position against the sun for a whole day (they would not be visible for more than a few hours after the sun rose) and they would not appear in the same position on the sun when seen simultaneously from widely separated locations such as Italy, Germany, and Poland—all contrary to what experience shows.[166] Whether the spots are planets is currently disputed and will be disputed for some time. Consult the writings of Apelles. Go to the *History* of Galileo.[167] Expect more in due time.[168]

DISQUISITION 31.

The blazes on the sun.

The blazes are little areas on the sun that are brighter than the rest of the solar body. Where and how such things may be will be told briefly at another time, but since the beginning they have been observed along with the spots.[169] Galileo writes the following regarding the blazes in the third letter of his *History*, page 132:

> On the very face of the Sun, one sometimes sees certain small areas that are brighter than the rest, and which careful observation reveals to have *the same motions as do the spots*. That these are on the surface itself of the Sun, I do not believe anyone can doubt, for it is in no way credible that there is some substance brighter than the Sun outside of it.[170]

DISQUISITION 32.

The elliptical sun.

The elliptical shape of the sun is nothing other than a contraction of the solar globe in height (along the direction of a vertical circle, such as BCD [in Figure 29-1]), especially noticeable at the horizon, even to the unaided eye. Over time, ignorance of this phenomenon has given birth to, nourished, and maintained many errors in astronomy and optics—errors that the acquisition of knowledge will take away. The elliptical sun is spawned by a *perpendicular* refraction that also explains many other things: why the sun appears higher than it should be; why the days are longer than they should be; why the relative positions of stars at the horizon are distorted; why stars linger at the horizon when rising or setting; and other phenomena.[171]

DISQUISITION 33.

The unevenness of the sun.

The unevenness of the sun appears at its limb, rendering the sun a most uneven circle. At one moment it is seen around the whole sun, then not at all, then on part of the sun. It is always unstable. It appears more often and more strongly in the morning and rarely in the evening. It is rarely seen by the unaided eye (only when it appears most strongly), yet it is frequently seen by the optic tube.

The one and only cause of this is the vapors present between us and the sun. These lie quite still over the lands throughout the night, but in the morning they are aroused by the warming rays of the sun. When heated, they rise. And, because they fluctuate and change in shape, because they are not perfectly diaphanous, because they carry water vapor in varying amounts, they agitate, cleave, mangle, and despoil amazingly the solar light that is passing through them toward our eyes. But as the day passes, the heating lifts these vapors higher, drives the water from them, and rarifies and homogenizes them, rendering them more stable. Thus the evening sun generally is more placid than the early morning sun.

The morning sun, especially after a nocturnal rain, emerges from behind the horizon, spreads along it, and then, as it advances, vacillates along its edge in and out so that its edge takes on the appearance of coarse teeth.[172] But if the weather has been most fine for some time, the sun will rise without such unevenness. From this it is certain that the unevenness of the sun must be attributed solely to the medium between it and us.

DISQUISITION 34.

Solar quivering.

A threefold trembling is noticed from the recently discovered spots. The first kind of trembling is a trembling of the whole solar globe and all conspicuous spots and blazes on it—a sort of vehement shaking about.

This we completely deny to be of the sun. It occurs in the morning often yet not always, in the evening rarely, and at noon most rarely, and at times it is absent for whole days. When the noonday sun here in Ingol-stadt is free from this trembling, other people at places either more to the west or more to the east see the sun trembling. If in a house you study one ray of sunlight that traverses the fumes rising from a hot stove, and a second ray that does not, you will find the first ray disturbed where it shines on a surface or passes through dust in the air, and the second ray not. Thus it is impossible for this trembling to be of the sun.

Rather, it proceeds from the antics of vapors. ‹Yes, vapors cause the solar trembling.› This is evident because the sun is never more than a little mangled one way or another. Indeed, the vapors—turbulent and swirling into pieces that blindly rush around and back into one another in different ways—block, weaken, and distort the solar rays, diverting them one way and another. So it happens that one part of the sun is seen to be heaved upward, another part downward, and the whole sun is twisted as though it were suffering from some digestive plague.

Experience supports this. If you observe the gilded globe of some tower from across an open plain on a morning when the vapors are peaceful, you will see the globe gleaming still and quiet in the sunlight.[173] But if you observe when the vapors have been stirred up by the rays of the sun, you will see each part of the globe differently, to the extent that the globe might appear to dance vigorously this way and that, up and down. What the vapors cause to happen in the globe, they also cause to happen in the sun, and in the stars, and so forth.

The second kind of trembling is a certain fluctuation of light that does not mangle or shake the sun. It is a certain boiling not unlike the twinkling seen in the fixed stars (and, to a lesser extent, in the wandering stars), especially when they are beheld by means of an optic tube with its eye glass removed [see disquisition 24]. We deny this behavior to belong to the sun or stars because these things occur in an unequal, irregular, and unordered manner. Rather, we argue that the cause of these things is again the vapors, acting in a twofold manner. As they swirl about between a celestial body and us, they may by means of refraction brighten its light in the usual manner or punitively strangle it.[174] These alternate rapidly so as to variously distort the rays of a celestial light, producing such varying gleams that those ignorant of this may be seized by wonder and easily led into astronomical or optical errors. The star Sirius, the great dog, repeatedly shows the signs of the brightening and strangling, being seen to dim wholly and then to brighten wholly. Terrestrial examples can also be found, as in the gilded spheres of the tower seen across the plain, which will be made to appear like gold sparkling with drops of morning dew, in a manner of speaking.

The third kind of trembling is observed not in every direction on the sun, but only along the direction of the sun's primary motion. It is a certain swaying advance of the solar image at some time or another. Thus a part of the sun might seem to not follow the ordinary progression of the sun across the sky, but might seem to stand still, or to advance too rapidly, or (rarely) to move backward.

This solar phenomenon we also attribute to the material between the sun and us producing refraction. There are two indications of this. One is that clouds passing before the sun will create a similar effect that then goes away when the clouds depart. The other is that this happens primarily just after the sun has risen or just before it will set, when it appears like a sudden unnatural stop or forward step. The cause of this is refraction and the disposition, shape, and number of vapors—as is explained in disquisition 32 on the elliptical sun—and as is supported by the repeated experience of astronomers and opticians (involving nothing new or unknown) and by simple observations of the edges of shadows projected by bodies.[175]

Apelles mentions two of these tremblings on page 44 of his second work. From Rome he writes that they are both nothing more than the scintillations of the sun, like the scintillations that are seen in the fixed and wandering stars—one more common, the other stronger and rarer—as he has discussed both recently and some years ago.[176] Nor have these troubles of the sun and stars escaped the notice of earlier writers. Indeed, Vitellius, in book 10, proposition 55, says, "Since light maintains the place of the image, in a body changeable by diverse motions, or by one strong motion, that form necessarily appears debilitated, stretched apart, and wrenched because of the movement of the underlying body [of vapors] in which it is seen," and further on he says, "And for this reason the sun always seems to scintillate for a while when beheld in the morning," and the form of it is disturbed "because it is received in disturbed airs" and a little later "thus therefore scintillation always happens to all fixed stars: since the cause of that is certainly the perpetual divarication of the form of a star at the place of the image, occurring as a result of the motion of the underlying body."[177] These discussions may not be entirely satisfactory to us, but nevertheless he acknowledges the effect.

Here then we are pleased to add the following corollaries.

DISQUISITION 35.

Corollaries.

The scintillation of the stars is not a result of their own spinning around.[178] It is not a result of their own internal fervor, or of their reflecting solar rays tremulously on account of their motions, or of their restlessly and irregularly discharging their own rays. It is not a visual tremor. It is not a sort of trembling of effort on account of the exertion needed to see. It is none of these, neither singly nor in combination. *The scintillation of the stars is solely the result of light, descending from them into the eye, being disturbed by the restlessness of the various intervening vapors.* ‹This is why the stars scintillate.›

Hence clear reasoning informs us:

Why all stars scintillate, but in different ways.
Why the fixed stars scintillate more, the wandering stars (that is, the planets) less.
Why stars lower in the sky do it more, while those higher do it less.
Why they do it more under a north wind, less under a south wind.
Why more when rain is near, less when the weather is fair.
Why the same star scintillates more when lower in the sky, less when higher.
Why it scintillates more today than it did yesterday, or will do tomorrow, when at the same elevation.
Why stars positioned near the zenith sometimes scintillate most vehemently for hours, other times not a bit for hours.
Why the small stars with feeble light scintillate more than the large ones with strong light.
Why stars scintillate more in one location on Earth than in another.

All the planets scintillate. ⟨Every one scintillates.⟩ Saturn does so with a gentle fluctuation, while Jupiter has gleams of brightness, but these scintillate less often and less strongly than the others. Mars scintillates especially when farthest from Earth. Venus does it most both when farthest from and when nearest to Earth, and less when in between. Mercury always scintillates. The moon, like the sun, is rarely seen by the unaided eye to scintillate.

You might deduce, not ineptly, that the light of the fixed stars that reaches us is much feebler in power than the light of the wandering stars—otherwise the planets would tremble more than the stars, for the reason that the planets would be more easily overpowered by the vapors. Some judge the light of Sirius, the Dog Star, to be more vigorous than the light of any planet ⟨(see Vitellius as cited previously)⟩. While its light does exceed Saturn, this is fallacious. The light of Sirius is very white, and for that reason it is more effective at provoking the eyes, but it is not simply more robust. Surely the splendor of the light of all the fixed stars is dilute and weak, while the splendor of the planets is full of vigor. *Who has ever read by the light of Sirius? Yet it may be possible to read by the light of Venus.*[179]

If we grant that light flows from the sun to illuminate the wandering stars, there is no reason why we should deny the same regarding the fixed stars.

Simon Marius, in the preface to his *The World of Jupiter,* indicates that *the solar tremblings occur not in the air but in the sun.* Those who contradict Marius do so correctly.[180] Indeed, he is mistaken in this, as he is mistaken about much else.

For example, he promotes the twinkling of the planets as something previously unknown, when in fact Vitellius (among others) long ago wrote in book 10 proposition 55,

> In the planets scintillation truly does not always happen, . . . unless by chance some great body of air, such as thick water vapor, is interposed between the vision and the forms of the planets. Then indeed on account of the uncertain motion of that vapor, the forms of the planets come to the vision as if scintillating. And from this cause sometimes we see the Sun itself scintillating in the morning. . . .[181]

And a little afterward he writes,

> In the planets truly scintillation happens rarely, because the cause of it happening is rare.[182]

Marius is also mistaken about whether the stars shine more strongly than the planets, and about whether any kind of trembling may be discovered in the moon. On the other hand, he has used the same method that we have used from the beginning, of viewing the stars by means of the convex glass [without the concave eye glass—see disquisition 24]. But he describes the spots seen in the bodies of stars observed in this way as being from the material of the glass. If he makes this assertion about all such spots, he is wrong. If he makes it about those that follow the glass when it is rotated, he is correct. The succession of colors that he describes seeing in Sirius certainly emanates chiefly from the eye.[183]

I shall briefly explain all these for clarity. When an optic tube with its convex glass is turned toward some star, two things are seen in the star: certain black voids and discrete bright regions. Of these, in turn, there are two kinds each.

There are certain very black little pools that are stable in position, number, and so on. These follow the glass if the glass is rotated. If the glass is changed, they change. There are also blackish little open spaces, like sectors of circles, moderately dark, separated by bright areas. These are not stable in place and shape for very long and do not follow if the glass is rotated. Rather, if the eye wanders, they wander with it, within the star. These are in the eye and from the eye.

There are also bright regions formed in the same manners as the dark regions. The light in them is made by the star, but they themselves are from the eye. Certain bright regions that arise and die away are generally little illuminated drops in the eye.

In regard to that vacillation of the sun as it progresses [that Marius discusses in his preface[184]—see also disquisition 34]: In truth it is noticed clearly and repeatedly in the border of light and shadow projected onto any wall. It has been noticed many years now, since before the invention of the optic tube. It can be seen by the unaided eye and detected daily.

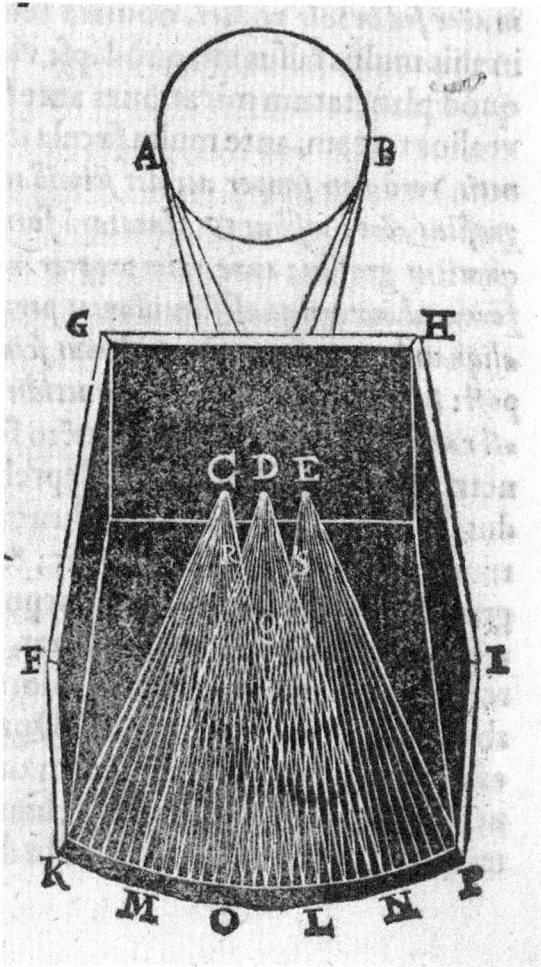

FIGURE 36-1

DISQUISITION 36.

A remarkable experiment concerning the light of the sun.

[Refer to Figure 36-1.] The sun AB, shining through identical holes C, D, and E into a dark room FGHI, sends three radiant cones, CKL, DMN, and EOP, of equal power on account of the equal holes, that illuminate a surface. Where these overlap there is a greater illumination. Thus segment OQL of the surface, illuminated by all three cones, is brightest. It is one and one-half times brighter than segment QRMO and three times brighter than segment RCKM (which in turn is half QRMO). The brightness of any segment illuminated by multiple overlapping cones compares to that illuminated by one single cone, as the number of cones compares to unity.

Inferences.

1. Rays of light and color are projected following straight lines.
2. Each point between the sun and the surface to be illuminated is a possible intersection point for the solar rays, from which they continue outward to form a radiant cone. For example, ACB becomes KCL, ADB becomes MDN, and AEB becomes OEP.
3. As the rays of the sun converge toward one another at the intersection points, likewise the cones diverge from those points.
4. As the object (the sun) is divisible, so are these radiant cones divisible. [A cloud that passes over a portion of the sun, the object, and divides it in two will be seen to pass over a portion of the three projected images and divide them likewise.][185]
5. Any point on the sun may radiate light in all directions—in a hemisphere. From this some might argue as follows: *As the object is divisible, so are the projected images or radiant cones divisible. Yet at least one object—a radiating point—is not divisible by any manner, and therefore neither is the radiant cone of that object. It may radiate a hemisphere, but that hemisphere is made up of rays extending from it and is not the point itself. Thus we may likewise regard the intersection point C: rays to it make up the object sun AB, rays from it make up the inverted image CKL. Point C is indivisible, but object AB is divisible, and so is its image KL. Therefore the object is divisible, but the images are not, for all possible images are in the indivisible point C.*

Response 1 (to inference 5). When it is said that as the object is divisible, so are the images, this refers to what is perceived. To divide the object in two is to divide [each of the three] images in two,[186] and this is observed. But to divide the image is not to divide the object. We deny the objection that the image is in the point C and say instead that point C is the vertex and common waypoint of the images. Visible images have length and width and are not indivisible. But if you are obstinate and contend that there are images in a point at C, we respond that they cannot be visible because nothing can be seen in a point.

Response 2. The visible object and the visible image define what is seen, not so much the vertex. Thus, one determines the other: the size of the object determines the size of the image, and the size of the image indicates the size of the object.

Response 3. Let us grant, for the purpose of argument (we do not, in fact, concede this), that a point may be visible, or that something may be visible inside a point. Such a point holds the same relation to the rays that converge into it, or diverge from it, as the point that is the center of a circle holds in relation to the semidiameters of that circle: it is the endpoint of those rays and thus will be virtually divisible, but not actually divisible. We might say instead that the point is not divisible, but the rays of it are divisible [for example, a semidiameter extended from the center of a circle to the left edge of the circle's circumference can be separated from one extended from the center to the lower edge, even if the point at the center is not divisible into leftmost and lowermost portions],[187] and this suffices. Or, finally, we might say it is divisible by degree, not by quantity. I add that the hemisphere of rays emitted from a point do not all fall into the eye. Only a few rays do, according to the capacity of the pupil. Of these, only one might be sensed. Thus, from one point comes one sensible ray, which imparts the image of a point.

6. Where the circles of illumination from the same object overlap on the same surface, there they light up that surface more. This is true whether or not the holes C, D, and E may be of equal size, and whether or not the light entering the holes is of equal intensity.

7. A similar thing occurs if someone uses illuminated colored things as objects, so that the light of each passes through the holes into the dark area and impinges upon an expanse of paper. There he will re-

gard these things in vivid color, each multiplied by the multiple holes, intersecting each other on the paper. From this it is evident that any one single point may send out a hemisphere of rays, and that same point will be seen multiply, in this place and that, by means of this ray or that. So what is seen is not only a matter of the nature of the things seen; what is seen in this case is light and color probably unchanged (as daily experience testifies), identical to what the *object being seen* imparted, and not directed, reflected, or refracted. Indeed, it is more than obvious that the visible appearances of things are the formed images of those seen things.

8. It is clear that the rays comprising the cones CKL, DMN, and so on are not all from the same point on the sun and do not number the same, but rather the rays of each cone are different from the rays of the other cones. Thus the same parts of the sun are indeed seen, but not in the same image.

9. Thus it is clear that if part of the breadth of an optical image is occulted, then a corresponding part of the breadth of the object of that occulted image is blocked. Conversely, if part of the visible object is blocked, then a corresponding part of the image will disappear.

DISQUISITION 37.

Experiment concerning fire.

An experiment like the previous ones can be performed at any time using flames. [Refer to Figure 37-1.] Equal flames F, G, and H are placed around the hollow cylinder ABCED so that they may shine equally through holes A, B, and C in the convex surface, onto one location DE on the opposite concave surface. The illumination of DE will be triple that of any one single radiant cone DAI, IBK, or KCE. Where the two cones DBE and DCE overlap at LKE, the illumination will be double that of a single cone like ECK, and two-thirds that of DLE. Granted cones of equal intensity, the place where they come together will be illuminated with an intensity proportional to the number of cones.

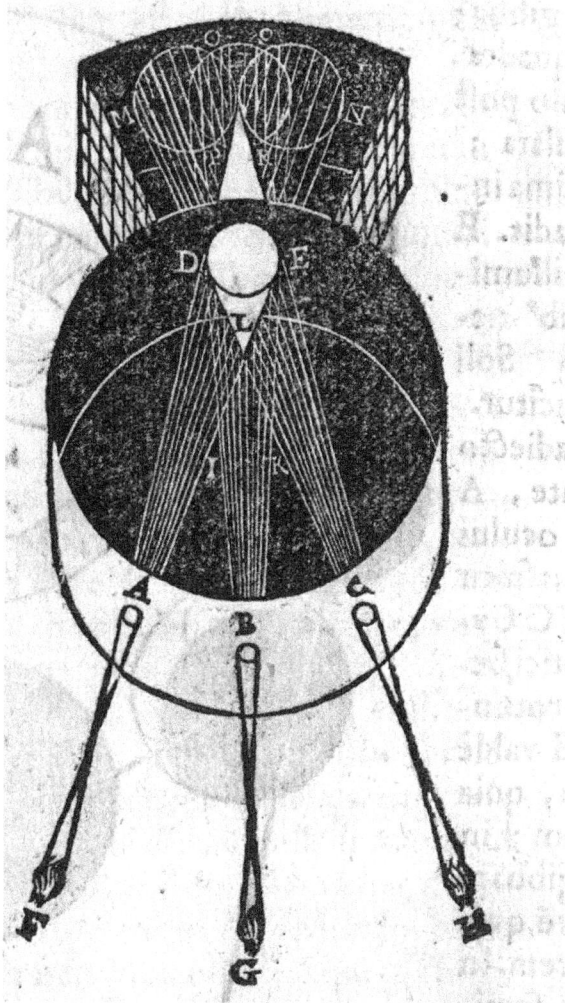

FIGURE 37-1

But if a hole is cut at that common location DE, the radiant cones will pass through and gradually diverge again from each other, and illuminate the opposite wall MN. They will form crescents MOP and QNR, whose concave sides face toward the doubly illuminated convex shapes PO and QR, and toward the triply illuminated area PR, where all the cones combine. Therefore, while the three cones do combine, they do not mix together. Otherwise, after mixing they would illuminate the wall equally.

DISQUISITION 38.

Concerning Venus.

Venus is revolved around the sun, according to the teaching of ancients and according to phenomena now recently discovered. Through the course of a year Venus displays the monthly phases of the moon—a most delightful spectacle. Indeed, when most distant from Earth, Venus displays the full phase, followed by gibbous, then the half-full phase when at its middle distance from Earth, then crescent a little later, and finally, when closest to Earth, it is dark and cannot be seen. Such illuminations necessitate that Venus circles around the sun.

[Refer to Figure 38-1.] The sun is A. The eye B on Earth beholds Venus by means of an optic tube. At C Venus appears round, but very small because it is at its most distant point from Earth. At D and E it appears gibbous, and larger because it is closer. At F and G—at quadrature—it is half-full. At this point it also produces its most brilliant light. At H and I it is crescent, and through the tube its size is comparable to that of the moon seen without the tube, if not larger. At K it is not at all visible because it is in conjunction with the sun and emits no light toward Earth. However, were Venus to pass in front of the sun, it would show as a great dark spot, for the diameter of Venus when closest to Earth is more than eight times its diameter when farthest from Earth.

This variety of appearances proves Venus to pass on both the far side and the near side of the sun. ‹A whole Venus is on the far side of the sun.› That it appears whole and very small proves it to pass on the far side. ‹A half Venus is on the near side of the sun.› That Venus is seen to be a dark body, half illuminated, half in shadow, demonstrates that it passes on the near side.

This is now shown. The eye B, line BL, and circle LFML (the exact boundary of the illuminated hemisphere of Venus LOM) are in the same plane. Thus arc LFM appears to the eye as a straight line. Straight FA, connecting centers F and A, stands at right angles to circle LFML. Therefore angle AFB is right, and triangle ABF is right. Therefore hypotenuse AB is greater than leg BF. Therefore when Venus is seen bisected—half illuminated, half in shadow—the center of Venus F is closer to eye B than is the center of the sun.

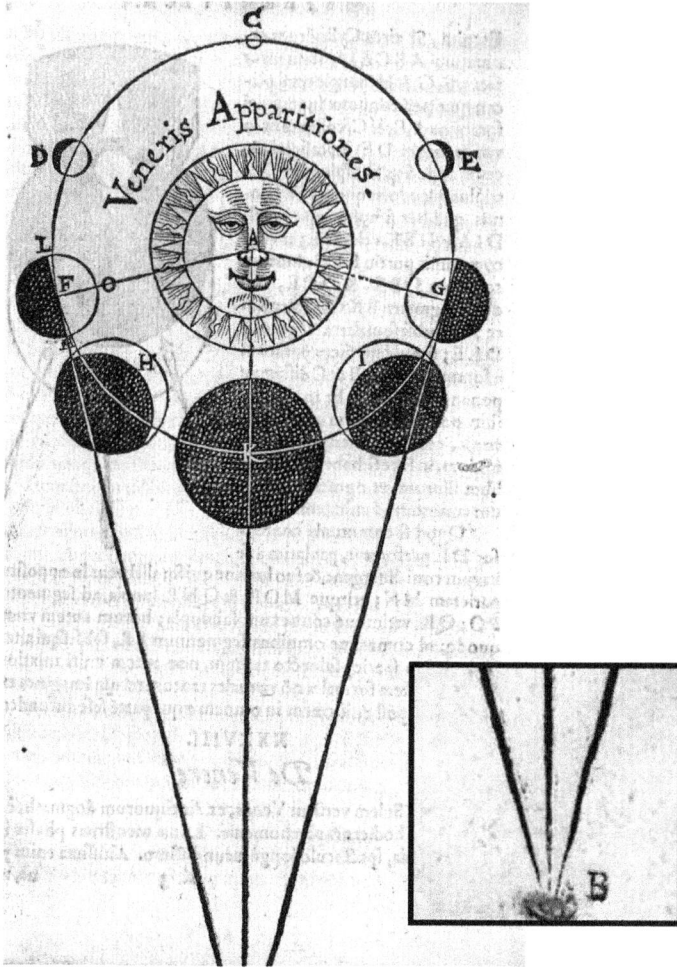

FIGURE 38-1

Mercury may also circle the sun. It is often seen to be spherical, and so it is nearly certain that it passes on the far side of the sun. Says Clavius in tome 3, chapter 1, page 42 of his book on Sacro Bosco:

The most ancient Egyptians, Plato in the *Timaeus*, Aristotle in *On the Heavens*, book 2, chapter 10, and *Meteorology*, chapter 4, have all supposed this to be the order in the celestial spheres—the moon might occupy the lowest place; the sun might immediately follow that; Mercury, that; Venus next. . . . Only in the little book to Alexander concerning the world has Aristotle (if nevertheless it is of himself) set Venus immediately above the sun, and beneath Mercury.[188]

Apuleus agrees with this in book 1 of his *On the World*,[189] as do some other ancients. Yet some assert Mercury to be crescent shaped. We have not yet experienced this ourselves and remain undecided. Whether what is definitely seen in Venus is also seen in Mercury—that as it approaches Earth it develops a distinct crescent shape, which little by little grows thinner and sharper until the horns of the crescent are most acute—is not easily settled by the optic tube. Therefore we reserve further discussion until a more exact examination is made.

But this phenomenon explains why Venus, when at C [in Figure 38-1] and very small, is blessed with a strong light nevertheless. It explains why when it is at quadrature, at F and G, it is most luminous as seen from Earth, for both its size and its light are increased. And finally why at H and I its light weakens. For while the luminous crescent is large in diameter, it is very thin, and thus the strength of its light is less than at quadrature.

Additionally, we make the following inference: the spheres of Venus and also Mercury being abolished, both of these wandering stars and the sun may be placed much closer to us, and therefore their sizes may be somewhat reduced.[190]

DISQUISITION 39.

Concerning Jupiter.

The remarkable company of Jupiter, first detected a few years ago by the outstanding, skilled, and learned Italian mathematician Galileo (although now this year, very much too late, a certain Calvinist[191] has endeavored in vain to persuade us otherwise), has seized the admiration of the whole school of astronomers, for four attendants or satellites, each different in motion, size, and distance, circle around Jupiter as though it were their lord.

The closest satellite is distant from the center of Jupiter by about 6 Jovian semidiameters, the next by 8 or a little more, the third by 10, more or less, and the farthest by nearly 20. ⟨The motions of the satellites of Jupiter are such that⟩ all move from west to east when on the far side of Jupiter, and from east to west when on the near side. They always appear smaller when they approach Jupiter from the west, and larger when from the east. They move faster when their separation from Jupiter is small, most slowly when at the extremity of their motions. They are eclipsed by Jupiter when moving toward the east, on the far side, never when moving toward the west. Therefore, sometimes they are closer than Jupiter and sometimes farther.

The satellites circle in a plane inclined such that currently they pass to the north when on the near side of Jupiter and to the south when on the far side. This is made apparent when they pass Jupiter or each other laterally, as seen in the observations recorded in [Figure 39-1]. The first of these (from 1612) was made in Rome and the rest here in Ingolstadt, except for the one labeled 14 March. In these the following should be noted:

1. The large circle in the middle is the wandering star Jupiter.
2. The line through the center of it is parallel to the ecliptic.
3. The small circles near this line are the companions of Jupiter, called the Medicean Stars by Galileo.
4. Those on the right are west of Jupiter; those on the left are east. Those above the line are to the north of Jupiter. They are on the near side of Jupiter, and thus they appear larger. Those below are to the south and on the far side. Those directly on the line are at their greatest distance from Jupiter, at the point where their motion stops and they reverse direction. At certain points they will pass so close to one another that they appear to form a single oblong star.

The last four illustrations [in Figure 39-1], for the night of April 29–30, clearly show how the satellites move over the space of six hours, from nine o'clock at night until three o'clock in the morning. It is very likely that the satellite that has appeared closest to Jupiter at hour eleven was previously hidden by the shadow of Jupiter, for it is not possible for it to have advanced so far in that time.[192] However, all these things may be better understood by means of another diagram.

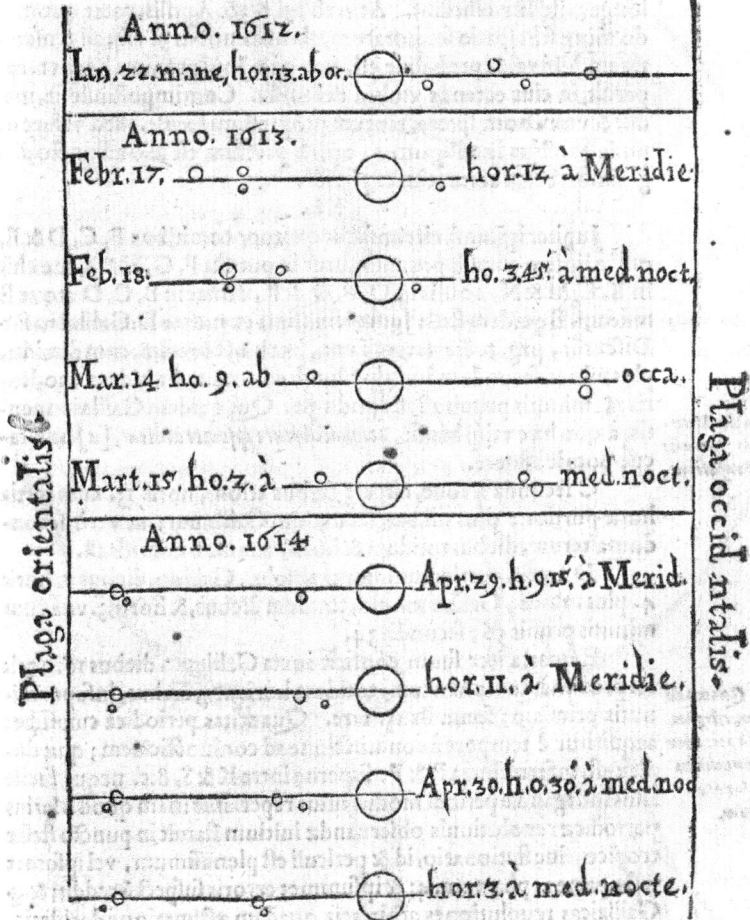

FIGURE 39-1

DISQUISITION 40.

[Refer now to Figure 40-1.] Jupiter is encircled by four companions B, C, D, and E. These revolve on their orbits toward points F, G, H, and I, respectively, on into K, L, M, and N, then into O, P, Q, and R, until they return into B, C, D, and E.

According to Galileo, on page 3 of his *Discourse*,[193] the wandering star closest to Jupiter, B, completes its circuit in 1 day and 18 1/2 hours, more or less. According to the author of *The World of Jupiter*, the period is 1 day, 18 hours, 28 minutes, and 30 seconds. (Having sponged off the discoveries of Galileo, he has merely improved the measurements, *as he himself seems to indicate* ‹in the preface of *The World of Jupiter*›.[194]) The second from Jupiter, C, goes around in 3 days, 13 1/3 hours, more or less, according to Galileo. According to the other guy, 3 days, 13 hours, 18 minutes. The third, D, completes its course in 7 days and 4 hours, more or less, according to Galileo. His emulator[195] gives this period as 7 days, 3 hours, 56 minutes, 34 seconds. The fourth, E, makes its journey in approximately 16 days, 18 hours, following Galileo; following Marius, 16 days, 18 hours, 9 minutes, 15 seconds.

‹How are the periods of the Jovian satellites determined?› The period of a given satellite is determined by measuring the time from one conjunction of it with Jupiter to the next conjunction. The conjunction can be on the near side of Jupiter, such as between B and F, or on the far side, such as between K and S. This is the most prudent method of determining the period. Marius has used the tropic or stationary point of a satellite to mark its period of revolution. This is perilous and prone to returning errors, as he himself testifies in part 2, phenomenon 4, of *The World of Jupiter*.[196] He may have determined the periods of revolution using a certain amount of conjecture and arbitrary estimation.

Indeed, when a satellite is near a stationary point, it may stand fixed in place, immobile, at least to the senses, for some hours or even days.[197] It is impossible to ascertain the true stationary point, except by chance.

On the other hand, the true midpoint of a conjunction can be found without sensible error, and from that can be found the period of the whole circuit of the satellite. For in a near-side conjunction, a satellite will be seen next to Jupiter both before and after the conjunction. The time halfway between these two events marks the midpoint of the conjunction. This can

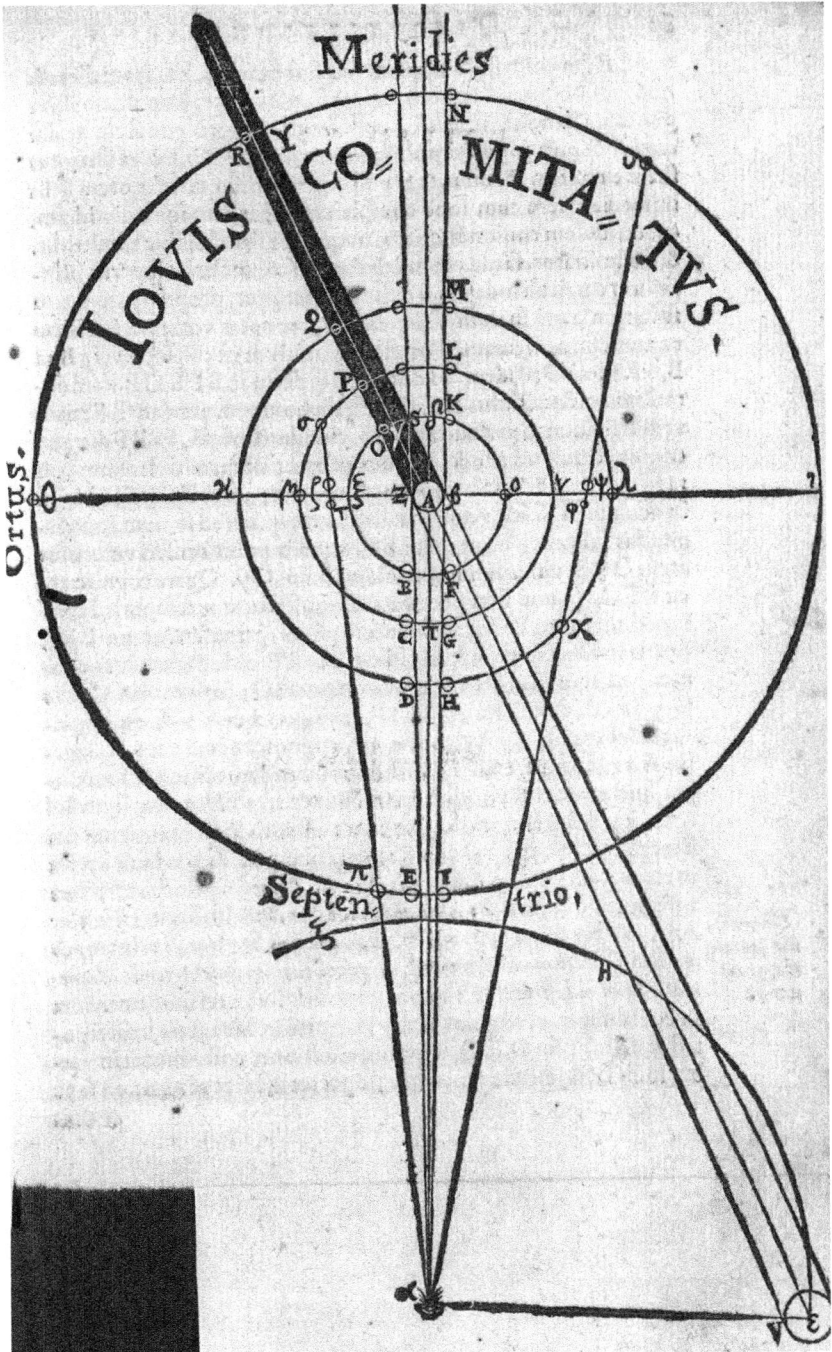

FIGURE 40-1. A letter X, to the right of the sun ε, is not visible in this image.

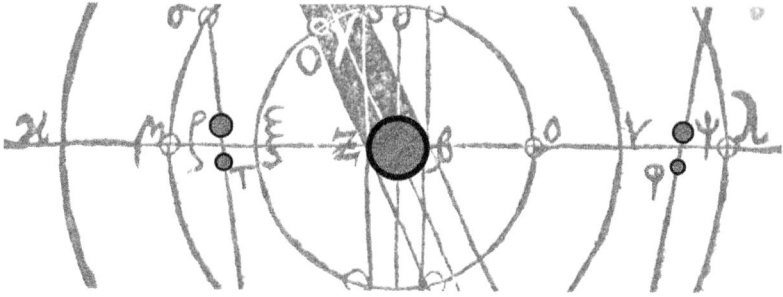

FIGURE 40-1A. In Locher's original book Figure 40-1 is a large, foldout illustration. This highlighted "seen from Earth" view, buried within the "seen from above the plane of the Jovian system" view, would thus be easier to see than it is in Figure 40-1. Note that the horizontal line is the θι line referenced in the text.

be determined with negligible error. If error is admitted, it can be corrected. In a far-side conjunction the times are more difficult to establish because of Jupiter's shadow and the apparent smallness of the satellites at that location. Also, the period can be measured from either B to B or F to F.

When I have the period BFKOB and the duration of the conjunction BF, I will easily have the arc BF, for the ratio of it to the duration of the conjunction BF is as the ratio of 360 to the period of the whole circle BFB. Arc BF may be supposed to equal the shaded arc OS, and half BF to equal Oγ (since rays VR and XY, sent from the sun and tangent to Jupiter, may be considered sensibly parallel on account of the great distance, as may rays Zα and βα, sent from Jupiter into the eye). Arc BFO may be determined by precise measurement of the time from first conjunction B all the way to the end of the eclipse at O. Arc BFδ is then known, from half of the period of the satellite's orbit added to half of BF. Arc BFδ removed from arc BFO leaves arc δO remaining. Remove from it arc Oγ (that is, half of arc BF) to leave behind arc γδ.

This arc will give angle γAδ, which is the same as angle αAε (the Earth-Jupiter-sun angle). Meanwhile, angle Aαε (the Jupiter-Earth-sun angle) is known from the positions of the sun and Jupiter in the heavens. Also known is the distance αε of the sun from Earth. With two angles and one side of triangle αAε all known, the entire triangle is fully known. Thus the distance of Jupiter from Earth Aα is known, the distance of the sun from Jupiter εA is known, and how much more distant Jupiter is than the sun is known.

⟨This is a Jovian phenomenon of great importance.⟩ *It is of such singular importance that it should be especially celebrated, and all astronomers should turn their efforts toward refining the observations of it.* The foundation of this whole thing is determining the angle δAγ. That in turn requires exact knowledge of the first emergences of the satellites from the shadow of Jupiter at O, P, Q, and R after they have been eclipsed. This will require diligent and frequent observations.[198]

Anyone who may be doubtful about the whole business of the eclipses may read Galileo's postscript to his third letter in his *History* concerning the solar spots, page 161.[199] Reason indicates that Jupiter is a dense body that reflects the light of the sun back to us and casts a cone of shadow. ⟨Yes, Jupiter casts a shadow.⟩ Experience shows this to be true. No other reasoning explains why, when Jupiter's satellites emerge from behind it, they make their first appearances separated from Jupiter by a large gap; why those appearances sometimes occur sooner, sometimes later; and why this happens only when they emerge from that region where the shadow lies, rather than whenever they are near their conjunctions with Jupiter.

DISQUISITION 41.

From these we have that Jupiter emits no light of its own—that in the manner of the moon it is an opaque body whose light is reflected solar light. Furthermore, the Jovian satellites are the same, or else they would shine when in the Jovian shadow. The satellites and shadow open a road to investigating the distance of Jupiter from Earth and from the sun, and also to investigating the true sizes of those bodies relative to each other. Indeed, perhaps from this shadow we may, among other things, confirm whether or not Jupiter is more distant than Mars.

What must not be disregarded here is that all the satellites of Jupiter are seen [in Figure 40-1] by the eye α to travel not in circles B, C, D, and E, but rather in the narrowest and longest ellipses. Sometimes they are on the diameter marked by θ and ι [see Figure 40-1a] and at other times almost on it. Indeed, they are on it itself when they occupy points θ and ι, κ and λ, μ and ν, and ξ and ο. They travel along the south side of the line θι—that is, below it as seen by the eye—when traversing the far side of their orbit circle. They travel along the north side, or above it, when traversing the near side, above the line when closer, below the line when farther.[200] Thus consider that if the fourth satellite is at π and the second satellite is at σ [in Figure 40-1], these will appear to the eye to be at ρ and τ (to the north and south), respectively [in Figure 40-1a]. Likewise, if the third satellite is at χ and the fourth at υ [in Figure 40-1], these will appear to the eye to be at ψ and φ, respectively [in Figure 40-1a].

But here arise some questions:

DISQUISITION 42.

Questions.

First—*Are these satellites in the same plane?*

Second—*Do their deviations from the line parallel to the ecliptic follow from Jupiter's deviation from the ecliptic,[201] so that it carries them with it, or do their deviations spring from some other source?*

Third—*Are their deviations all the same, and if not, then by how much do they differ?*

Fourth—*How does it happen that they appear generally larger when they seem more separated from Jupiter (near their stationary points) than when they seem to draw close to Jupiter?*

Fifth—*Why are the satellites of Jupiter seen better (all else equal) in twilight than in the dead of night?* The same must be asked concerning Venus and the other planets.

Sixth—*How does it happen that a Jovian satellite generally appears smaller and darker and more difficult to see when Jupiter is excluded from the field of view than when Jupiter is viewed together with the satellite?*

Experience will answer the first and second questions by the year 1617. During June, July, and August of that year Jupiter will be very close to the ecliptic.[202] If at that time the satellites coincide entirely with the ecliptic, and eclipse one another, and are fully eclipsed by Jupiter, then it will be clear that they are all together in the ecliptic plane. But if at that time they are seen to stray outside the line θι [as in Figure 40-1a], then it will be certain that the plane of their orbital circles deviates from that of the ecliptic.

In the meantime, it seems probable that they all circle in the same plane, but deviate from the line θι on account of Jupiter. This is because, first, those which are more distant from Jupiter deviate more, while those which are closer deviate less; and second, it seems probable that the satellites follow Jupiter in latitude just as they do in longitude. Therefore, after 1617 their deviations would be reversed, such that they are to the south of Jupiter when on its near side and to the north when on its far side. But time will tell.

The answer to the third question is that the deviations are all different. The first satellite (nearest to Jupiter) deviates from line θι least of all, the second deviates less than the third, and the third less than the fourth. The reason is that they all are in the same plane, which is inclined to the eye. Thus those satellites that are more remote from the center of Jupiter are seen to be separated from the center by a greater angle. This angle, however, has never been seen to exceed the semidiameter of Jupiter. From this, if we state the diameter of Jupiter to be 2 & 3/4 minutes of arc, following Tycho,[203] the fourth or outermost satellite deviates from line θι by about one minute, the third by half that, the second by twenty seconds, and the first by fifteen, more or less. Hence, all else being equal, the first satellite passes through the greatest amount of Jupiter's shadow, while the rest pass through progressively less.

In regard to the fourth question, it is easier to refute the causes given by others than to give our own cause. That the cause is a sphere of vapors around Jupiter[204] is a wishful fiction. Why would satellites be diminished by such a thing when on the near side of Jupiter and thus in front of such vapors? This is especially true in regard to the outermost satellite, which is so far from Jupiter. Can the vaporous sphere be so large as to include that outer satellite? Moreover, that sphere would be either more or less dense than the surrounding regions. If less, then it would have little effect. If more, then it would refract the light of satellites as their light passed from denser material into more tenuous material (according to the law of refraction) and, if anything, make them appear larger. Therefore this reasoning is inane.

A certain person alleges that the cause of this might be that the satellites are also illuminated by Jupiter.[205] Thus they will show little phases to us, like the moon. But what then? Wherever a satellite may be situated around Jupiter, it will show to our eyes a full spherical face, illuminated by the sun. When it is situated so that it may be illuminated by Jupiter, the face illuminated by Jupiter is not turned toward our eyes.[206] It goes without saying that it is stupid to suppose that this weak, secondary, supposed illumination could be perceptible in so small a body at so great a distance.

And by setting aside good reasoning here this person does great injury to himself in the view of those who are learned with regard to the world of Jupiter. No one buys a product when he knows it to be poorly made; thus no one will buy what this vendor is selling. His calculations regarding the Jovian company do not agree with the heavenly phenomena, however much you test them. But if you do not have confidence to test by observations, you may consider what is provided here, produced with confidence and diligence, not by just one person skilled in these things but by many. You will recognize that the calculations of Marius do not stand.

But to return to the matter, the principal cause [for why the Jovian satellites appear larger when they are seen more separated from Jupiter (near their stationary points) than when they seem to draw close to Jupiter] is the Jovian torch itself. Its brilliance obscures the poor satellites, especially those nearby. This is a strong argument because even when their orbits carry them on the near side of Jupiter, satellites appear reduced when close to Jupiter, and when very close they disappear to a certain degree (unless you move the optic tube to see the satellite alone, without Jupiter). Contributory causes are better and worse quality of air, the altitude of Jupiter above the horizon, and the differing aptitude of the eye. However, anyone who can see has seen a lesser light obscured by a greater, a weaker by a stronger.

The answer to the fifth question is that it happens for the same reason — because under the light of twilight the eyes are not overwhelmed, and vision is stronger. Experience with the crescent Venus supports this much; Venus is observed more clearly and distinctly under some twilight than under dark of night. Moreover, when Jupiter is near its farthest point from Earth, the effect is lessened, either because Jupiter is less bright or because the satellites appear smaller, and they require night to be seen.

The answer to the sixth is that the light from Jupiter illuminates the eye and is conducive to vision. This is not opposed to what is stated regarding question four. Just as the glare of Jupiter overwhelms the eye and greatly diminishes the appearance of the satellites, so the light it casts out broadly accentuates those lesser lights and causes them to be distinguished better.

Thus practical rules for observing Jupiter are:

1. Two or more Jovian satellites visible in a straight line through the center of Jupiter, unless one is to the east and the other to the west of it, are quasi stationary and at their maximum separation from Jupiter.
2. A satellite which is above a line passing through the center of Jupiter and three other satellites is orbiting on the near side of Jupiter, is to the north of Jupiter, and moves from east into the west until it reaches its stationary point.
3. A satellite which is below a line passing through the center of Jupiter and three other satellites is orbiting on the far side, to the south, and moving from west into the east.
4. Every satellite moving from east into the west is on the near side of Jupiter and to the north.
5. Every satellite moving from west into the east is on the far side of Jupiter and to the south.
6. Satellites that appear brighter than the others generally are on the near side of Jupiter and move from the east toward the west.
7. No Jovian satellite can pass through the center of Jupiter, nor pass through the center of another satellite.
8. It is possible that Jovian satellites will at some time form a straight line with the center of Jupiter following east to west.[207] They may form a line only to the east of Jupiter, or only to the west, or both. This line will not necessarily be either coincident with or parallel to the ecliptic.
9. It is possible that the satellites may assemble into one straight line following north to south, which cuts at right angles the line above.[208] In this rare concurrence it will not be difficult to compare the diametral magnitudes of the satellites with the diameter of Jupiter.
10. The Jovian satellites are able to form a variety of triangles and rectangles with right, obtuse, and acute angles.
11. They form diverse quadrangles of diverse appearances, and so on.

DISQUISITION 43.

Different moons.

Until this time we have acknowledged one moon, circling around Earth, which the unaided eye A [in Figure 43-1] positioned on Earth sees dark when at B, half when at C, full at D, and short of full at E. At times it appears equal in size to the sun F (from which it receives all light), at other times a little smaller or larger.

But truly with the discovery of the Astronomical Eye (the optic tube), more moons have revealed themselves to us. Venus is one especially. Indeed, Venus viewed by the eye A, through the tube GH, appears crescent when at I, grows as it moves away from M, becomes full at K, and half-full at L (when most separated from the sun). At M it appears entirely dark. *And thus we have circling the sun a second moon with an approximately annual cycle.*

As regards Mercury, nothing should be said. Solid experience is needed. Yet others say Mercury is the same as Venus, and this is not improbable.

We now have an additional four other little Jovian moons, detected not by the eye, nor by the tube, but by reason. Since they are in fact eclipsed by the shadow of Jupiter, they will present the same appearances to Jupiter that the moon or Venus present to Earth. And so an eye positioned on Jupiter might see a satellite crescent shaped when at O, full when at P, half at Q, and dark at R. Moreover, one of these little moons completes a cycle in about one and one-half days, another in about three days, the next in about eight days, and the last in about seventeen[209] days.

If a man were placed on Saturn, he might observe all the planets to be images of moons. And the number of these might need to include the solar spots, if they be wandering stars or permanent bodies (as some consider, but this is disputed and a matter of vigorous inquiry).

DISQUISITION 44.

Concerning Saturn.

Saturn still turns astronomers upside down or truly toys with them, because of either spite or malice. Indeed, ‹such are the phenomena of Saturn

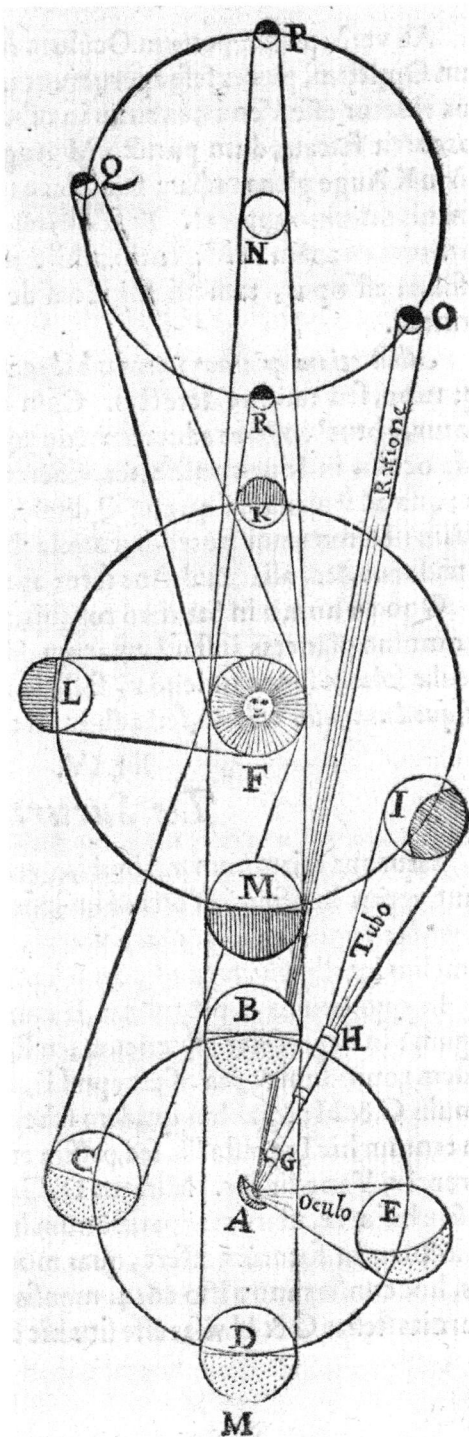

FIGURE 43-1. The gibbous moon E shows how the moon appears "to the eye" (Oculo) when at that location on its orbit around Earth. The crescent Venus I shows how Venus appears "to the tube" (Tubo), or telescope, when at that location in its orbit around the sun. The crescent Jovian moon O shows how according "to reason" (Ratione) that moon must appear as seen from Jupiter when at that location in its orbit around Jupiter.

that⟩ it presents different visions to them—*it appears now single, now three-fold; sometimes elongated, other times round.* This will be understood better by reference to [Figure 44-1]. In this figure the eye A gazes through the optic tube B at Saturn, beholding it oblong like C, perfectly round like D, or threefold like F, G, and H (here Saturn itself is round F, but accompanied by two attendants G and H). Indeed, it has presented itself with such diverse shapes not only here at Ingolstadt, but in Italy at Rome, Florence, Venice, and everywhere. Galileo, toward the end of his third letter in his *History,* marvels at the inconstancy of those attendants and makes some conjectures about the future,[210] which we do not call into judgment. This figure corresponds to February 12, 13, and 14 of the year 1614. The Saturnian stars G and H were seen in the position delineated here: G indeed appeared brighter and farther from Saturn than H.

We shall not facilely ascribe this to some deception of the optic tube or of the eye. Indeed, it is certain that it cannot be so ascribed, as Galileo testifies. But then the important question arises: how does this fickleness occur? It must occur either because of movement from place to place or because these associates of Saturn actually vanish. So far no one has claimed the latter. But if this occurs entirely because of movement, what is that movement? ⟨These are some of the phenomena of Saturn requiring explanation.⟩ Astronomers are perplexed.

Indeed, these Saturnian stars either revolve around Saturn by means of their own motions or are dragged around by its motion. If the first is the case, then it is necessary for them to approach to Saturn, to recede from it, to be occulted, and so on. These things have not been clearly observed so far, but what has been observed might be vague evidence for them. Saturn is seen solitary at one time, obviously when the attendant stars might stand behind it or before it. At another time it is seen in the form of an egg, when they might be conjoining to or separating from it. At yet another time Saturn appears triform, clearly when the stars G and H might be near their stationary points.

If the second is the case, it ought to occur by one of two methods. Saturn might itself rotate around its own axis, and wheel those stars around with it, and hide them in conjunction. This method might be possible, but it is not very satisfying, for it depends upon a kind of conjecture only. The other method is that those stars might be carried by the annual mo-

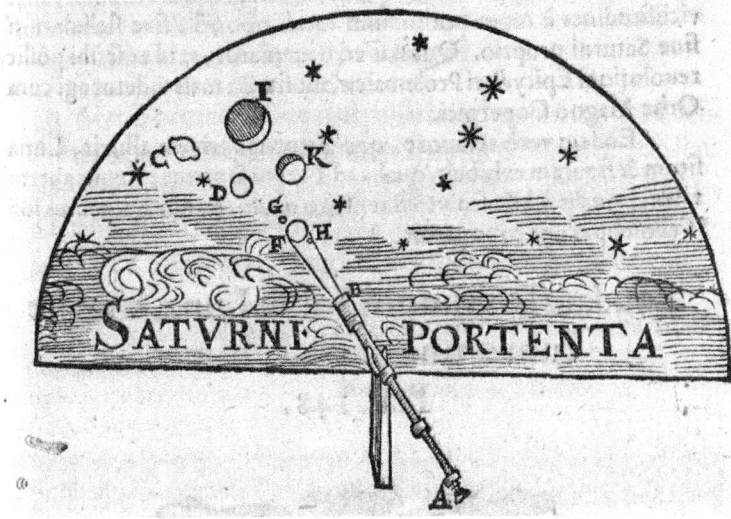

FIGURE 44-1

tion of Saturn's epicycle, together with Saturn in a fixed arrangement, moving with Saturn in locked step like slaves.

To find out what actually happens and settle these matters, the breadth of Saturn must be diligently examined to determine when it is greater and when it is lesser.[211] But we suspend judgment for now as regards all these matters of Saturn, and leave them to be decided by further experience with the phenomena. We make only this judgment: these vicissitudes of Saturn arise chiefly on account of motion from place to place, whether that be motion of the little Saturnian stars themselves or motion of Saturn proper. But if it may be determined at some point that this motion can be ascribed to the revolution of a Ptolemaic epicycle, or the like, it will go badly for the Great Orb of Copernicus.[212]

At the time when Saturn appeared triple, the moon displayed to the free eye the position and shape seen at I in [Figure 44-1]. Venus showed the position and shape seen at K, visible only to the telescope[213] or tube.[214]

PRAISE YE HIM, O SUN AND MOON:
PRAISE HIM, ALL YE STARS AND LIGHT. PSALM 148[215]

Approval

By these astronomical disquisitions the new and remarkable phenomena of the planets are diligently observed, their causes cleverly investigated and plainly explicated, and for this reason the disquisitions are deserving to be delivered to the press and led forward from the shadows into the sun and into the arena.

Johannes MocQuetius of the Society of Jesus,
Professor of Sacred Theology
and Dean pro tempore of the College of Theology.

I decree these disquisitions to be indeed commendable in novelty, and suitable for printing and public disputation.

Simon Felix of the Society of Jesus,
ordinary Professor of Philosophy
and Dean at this time of the College of Philosophy.

A Poem to the most Noble and Learned Lord

First you begin to commit yourself to the starry battle, and thus you show the learned flow of your mind. Honor is begotten by the battle; it does not beget it. You forsake the praises that merely embellish the first glory of study. You delve for years into the secret chambers of the mathematicians; learning the doctrines is your attentive concern. All doctrines are penetrable to virtue. Virtue begets the battle. The way is known to you, and you shall not fall from it. You honor heaven by your talent and Earth by your character. Light has come to both by your rays. Certainly it is common for great minds to comprehend heaven and Earth through talent and character.

Joannes Dornwogginger,
Bavarian of Ingolstadt,
Bachelor of Arts and of Philosophy,
Candidate for Master, Student of Law.

A Poem to the most Noble and Learned Lord

The deep secrets of nature might lie hidden, but the most precious thing is worth the struggle to bring it forth. For this reason men probe deep into the unknown to find cures for illness or sources of legendary magic. For this reason eagles of legend soar to the stars. Now a new winged bird has brought the stars to the eagles. Oh nature, he has revealed what tricks you have! You place voids in the middle of the sun! You make stars that cast shadows behind them and carry torches around them! And when we seek to scrutinize the most splendid of your works of art, we find that the sun radiates its splendor with blazes beside SPOTS!

Corbinianus Abentheurer,
Bavarian of Freising,
Bachelor of Arts and of Philosophy,
Candidate of Master, Student of Theology.

(Here Locher includes a listing of sixteen typographical errors discovered in the printing.)

THE END

NOTES TO THE TRANSLATION

1. These are references to Orpheus, a poet and musician from Greek mythology whose music had the power to move inanimate objects, who voyaged with the mythological Jason and his crew of sailors in the ship *Argo* to search for the Golden Fleece, and who attempted to rescue his wife Euridice from Hades.

2. A giant with one hundred eyes (from Greek mythology).

3. Possevino (1533–1611) was a Jesuit theologian and papal envoy.

4. Possevinus 1593, 173.

5. For the long quotation that follows, see Possevinus 1593, 173–78. Locher has marginal indices for each of the topics mentioned by Possevino: philosophy, theology, medicine, agriculture, mechanics, military, law. These are not reproduced here.

6. Book 1, chapter 2, line 185a17; Aristotle 1984, 1:316.

7. Book 2, chapter 9, line 200a15; Aristotle 1984, 1:341.

8. Book 1, chapter 1; Aristotle 1984, 1:114–15.

9. See, for example, book 3, chapters 4 and 8; Aristotle 1984, 1:496, 501–2.

10. See, for example, book 3, chapters 3 and 5; Aristotle 1984, 1:600–601, 604–6.

11. The Council of Nicaea, convened in the year 325.

12. The Council of Trent, convened in the year 1545. Locher does not name Trent specifically, but rather refers to the most recent synod.

13. Jewish king, son of David, tenth century BC.

14. Greek physician and writer on medicine (ca. 130–ca. 200).

15. Greek biographer (ca. 46–ca. 120).

16. Archytas (ca. 430–ca. 350 BC) and Eudoxus (ca. 390–ca. 340 BC) were Greek mathematicians, astronomers, and philosophers.

17. See North 1898, 276.

18. Archytas is reputed to have created a flying mechanical bird, possibly powered by a jet of steam.

19. Posidonius of Greece (135–51 BC).

20. Roman statesman and orator (106–43 BC).

21. Cicero 1884, book 2, sections 34–35, 71–72.

22. Line 36d; see Plato 1961, 1,166.

23. Stories of this device, citing Possevino, were circulated even in the nineteenth century. See, for example, Wanley 1806, 383; *Scientific American* 1896, 250.

24. *Republic*, book 7, line 526d; Plato 1961, 759.

25. Roman general and consul (ca. 268–208 BC).

26. One of three giants from Greek mythology who had fifty heads and one hundred arms. Also called Aegaeon.

27. Translation of Marcellus's remarks from North 1898, 281. Locher provides a brief paraphrase, not this exact quotation.

28. John Zonaras, a twelfth-century Byzantine chronicler and theologian.

29. Proclus of Greece (411–485).

30. Quotation attributed to Zonaras in Wesley 1823, 388–89. Locher provides a brief paraphrase, not this exact quotation.

31. Clavius (1538–1612) was a leading Jesuit astronomer and mathematician.

32. Johannes de Sacro Bosco (ca. 1195–ca. 1256) wrote a treatise on astronomy that was widely referenced for centuries.

33. See Proclus 1792, 62–65.

34. The Latin that Locher quotes can be found in, for example, Proclus 1560, 21–22. The English translation I have borrowed from Proclus 1792, 74; the italics are my addition.

35. See Figure N-1 for an example of how Locher uses a marginal note to cite Proclus here.

36. See Proclus 1560, 22; English from Proclus 1792, 76.

37. See Proclus 1560, 4; English from Proclus 1792, 50.

38. See Hart 1959, 27, 174–76, 219. In this paragraph and the following paragraph in particular, and in disquisitions 2 through 7 in general, Locher relies heavily on the technical language of Aristotelian metaphysics for his discussion: "substance," "accidents," "predicaments," and so forth. It is likely that more readers of this translation will be well-versed in vector calculus than in Aristotelian metaphysics, and therefore to most readers even a perfect rendition of Locher's words would be less meaningful than would be an excursion into calculus. Moreover, *Mathematical Disquisitions* is not a book on Aristotelian metaphysics, and these disquisitions are short and preliminary. Locher most likely did not intend that his readers become bogged down in these disquisitions, but rather he expected readers to have sufficient familiarity with metaphysical language to be able to quickly read through them. Thus, throughout disquisitions 2 to 7 I have endeavored to render

mouetur, &c.

Obiectum itaq; Mathematicæ secundum hos, non sola Quantitas, sed Numerus, etiã siue in abstracto siue in cõcreto sumpt°, &c.

Procl. l. , Rursus autem quidam alio modo diuidendam esse Mathema-
eod. c. 13. , ticam censent, sicuti & Geminus, &c. Eius enim quæ in intellectili-
, bus versatur duas longè primas ponunt partes, *Arithmeticam &*
, *Geometriam*; eius verò quæ in sensilibus officium & opus explicat
, suum, sex: *Mechanicam*; *Astrologiam*; *Perspectiuam*; *Gœodæsiam*;
Canonicam atq; *Supputatricem*, &c.

Procl. l.   Idem Proclus alibi ait, considerari à Mathematico, Propor-
eod. c. 3. , tiones, earumq; Compositiones, Diuisiones, Conuersiones, & al-

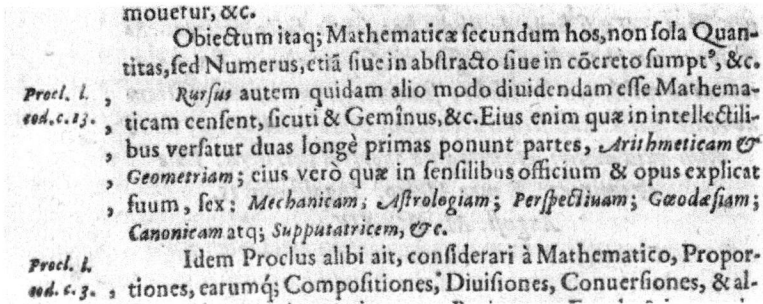

FIGURE N-1. Illustration of how Locher cites his sources. The lines preceded by a ' are taken from Proclus, who is acknowledged in the marginal note at the left. In the translation these marginal notes are worked directly into the text and ‹indicated thus›.

Locher's discussion loosely and briefly, and even to paraphrase. My purpose is to convey to modern readers who are unfamiliar with Aristotle some sense of what Locher is talking about, without bogging them down in metaphysical language. I myself am more familiar with vector calculus than Aristotelian metaphysics, so I have relied on a metaphysics text (Hart 1959) to aid in this regard. Hart 1959, 174–76 pertains to the subject of "accidents," the term Locher uses here with regard to quantity. Hart 1959, 27 pertains to quantity conceived generally, being almost an entity itself. Hart 1959, 219 pertains to extension.

39. See Hart 1959, 27, 216–17 for a discussion of quantity being "discrete" or "continuous." There are some difficult sentences here. See the previous note regarding the language of metaphysics. Locher distinguishes the continuous whole as being either "pervias aut impervias," which I translate directly as "per vias"— "through roads"—following Hart 1959, 216–17, regarding continuous quantities being successive (where there is a definite path, or order of before and after, like time) and not successive (space can be divided up without a definite before and after). Thus the clarifications in brackets refer to Hart 1959, 216–17. The identification with arithmetic follows Hart 1959, 27.

40. "Things which coincide with one another are equal to one another" (Euclid 1956, 1:155), or "Magnitudes which exactly fill the same space are equal to one another" (Hawtrey 1874, 6).

41. This sentence is difficult. The radical quantity to which Locher refers may be infinity, which he will discuss shortly.

42. The bracketed phrase is not in Locher's text directly. However, knowledge of this concept seems presumed in order for the Proclus quotations within this disquisition to make sense. Therefore I have added the phrase so the reader who is not familiar with this will understand what Proclus means by "supreme feat." See Hart 1959, 25–27 on the "supreme feat."

43. Proclus 1560, 18; Proclus 1792, 71.

44. Proclus 1560, 18; Proclus 1792, 70–71.

45. The Proclus reference is possibly to book 1, chapter 6—Proclus 1792, 55—which speaks of mathematical things being "the genuine offspring of the soul."

46. Here Locher has a footnote listing various specific kings, with a reference to Clavius.

47. Locher uses the term *universum* to describe both the observable universe of Earth, sun, moon, planets, and stars (he also uses the term *mundum*—"world"—for this) and the unobservable, supposed broader chaotic structure that produces universes. *Multiverse* is a modern term that well fits this broader structure. For additional discussion on the idea of a multiverse in the writings of the seventeenth century, see Danielson 2014, 27–50, and especially 38–39, where *Disquisitions* is specifically considered.

48. Locher uses the term *atomus*—not the modern idea of an atom, but an elementary particle that cannot be further divided.

49. Book 1, proposition 26: "If two triangles have the two angles equal to two angles respectively, and one side equal to one side, namely, either the side adjoining the equal angles, or that subtending one of the equal angles, they will also have the remaining sides equal to the remaining sides, and the remaining angle to the remaining angle." Book 1, proposition 29: "A straight line falling on parallel straight lines makes the alternate angles equal to one another, the exterior angle equal to the interior and opposite angle, and the interior angles on the same side equal to two right angles." See Euclid 1956, 1:301, 311.

50. Book 6, definition 4: "The height of any figure is the perpendicular drawn from the vertex to the base." See Euclid 1956, 2:188.

51. Book 1, proposition 11 discusses how to draw a straight line perpendicular to another straight line. See Euclid 1956, 1:269.

52. Book 6, proposition 8: "If in a right-angled triangle a perpendicular be drawn from the right angle to the base, the triangles adjoining the perpendicular are similar both to the whole and to one another." See Euclid 1956, 2:209.

53. Book 1, proposition 19: "In any triangle the greater angle is subtended by the greater side." See Euclid 1956, 1:284. Angle GAF is a right angle, subtended by side GF. Angle AGF is supposedly less than a right angle, subtended by side AF.

54. While Copernicans retained the Aristotelian physics that described motion on Earth in terms of four elements—namely earth, water, air, and fire—that had natural motions directly toward (in the case of the element earth) or away (in the case of the element fire) from Earth's center, they differed from Aristotle in that the center of Earth was not the central, or lowest, place in the universe. Thus the realm of Aristotelian physics was contained within the orbit of the moon and circled the sun as a whole. See Figure N-2.

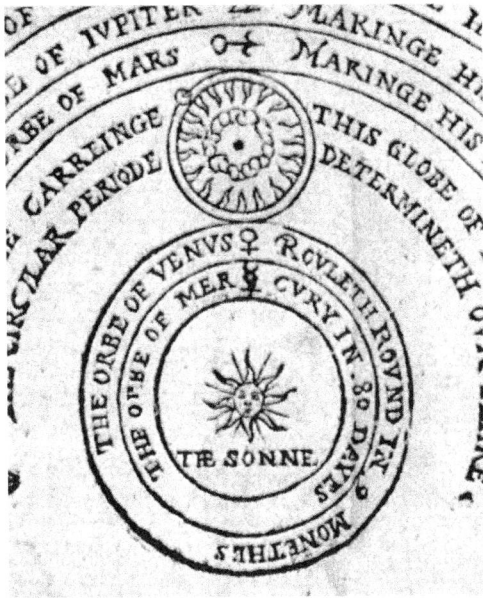

FIGURE N-2. Detail from Thomas Digges's 1576 "Perfit Description" illustration of the Copernican system, showing the "sphere of the elements" traveling around the sun with Earth. Elemental fire is just within the sphere of the moon, with air below it, and Earth itself at the center. Within this sphere is the realm of Aristotelian physics and elements. Heavy objects move toward its center, while fire rises toward its circumference. Image courtesy History of Science Collections, University of Oklahoma Libraries.

55. All bracketed remarks have been added for clarity. Today the fact that Earth's axis maintains its orientation toward the North Star is not considered to be an additional motion of Earth, but rather a natural consequence of the conservation of angular momentum—a natural gyroscopic behavior. In Locher's time, however, it was considered a third motion. Indeed, replicating this phenomenon on a mechanical model or orrery (see Figure N-3) requires a specific mechanism. Such an orrery thus requires an arm to swing the model Earth around its sun, a second mechanism to rotate the model Earth, and a third mechanism to maintain the direction of the model's axis; otherwise the axis would maintain its orientation relative to the model sun.

56. Here Locher is referring to Jesus descending into hell following his death on the cross. Since he died in the middle of the day (see, for example, Luke 23:44–49), if hell is presumed to be at the center of Earth, then Jesus's motion toward hell was motion generally away from the sun. Since the sun is at the center of the Copernican universe (the lowest point, as seen in Aristotelian terms), this would imply that his motion was "upward."

57. Here Locher is referring to the ascension of Jesus into heaven (Acts 1:6–11). As the ascension presumably occurred during the daytime, Jesus's motion up from Earth was motion in the general direction of the sun. Since the sun is at the center of the Copernican universe (the lowest point, as seen in Aristotelian terms), this would imply his motion was "downward."

FIGURE N-3. A mechanical orrery *(above)* that demonstrates the motion of Earth. One mechanism moves the model Earth around the sun, another rotates Earth, and a third *(below, detail)* keeps the axis of Earth pointed in the same direction in space. Images courtesy of Todd Timberlake, Berry College.

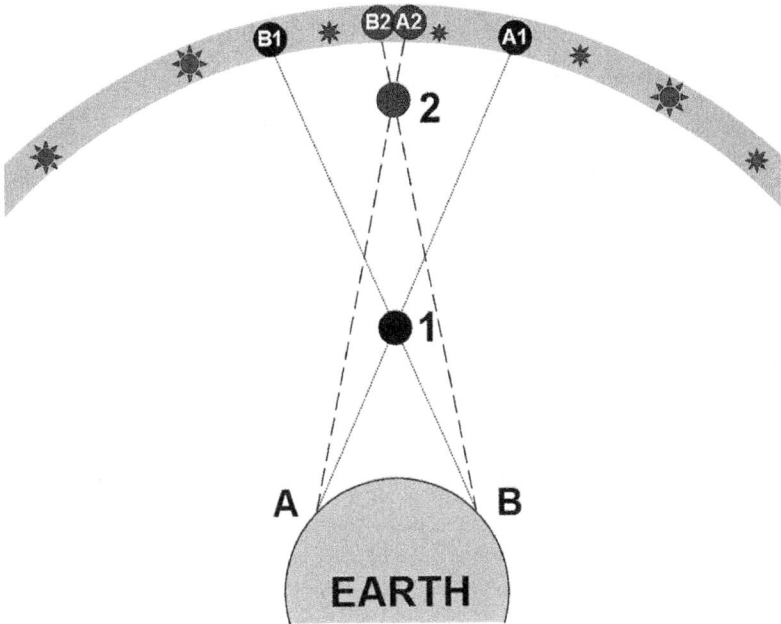

FIGURE N-4. Object 1 appears among the stars at A1 as seen by an observer on Earth at A, but it appears at B1 as seen by an observer at B. The difference between the two positions is the *diurnal parallax* of the object. Object 2 is more distant from Earth and thus has a smaller diurnal parallax—there is a lesser difference in the positions of object 2 as seen by the two observers.

58. Galileo includes this entire paragraph, and the first sentence of the next paragraph (about the universe being inverted), in his *Dialogue*. See Galilei 2001, 415.

59. Copernicus 1995, 14–15, 27.

60. Most astronomers in Locher's time accepted that the Earth–sun distance was roughly 1,200 terrestrial semidiameters. For more on the 1,208 value, and the question of the sun's distance, see Van Helden 1985, 48. Locher cites Maestlin's appendix to Georg Joachim Rheticus's *Narratio Prima* for this value (and for the measurements in German miles; a German mile is a little less than five U.S. miles); see Kepler 1596, 168.

61. Diurnal parallax is a difference in position (relative to the stars) of a celestial object when seen from different locations on Earth (or from the same location at two different times). See Figure N-4. Objects that are sufficiently far from Earth have no measurable diurnal parallax; the size of Earth is negligible compared to the

*A*RGVMENTVM EX MOTV ET
ORBE ANNVO.

In fententia Copernici omniumq; Copernicanorū inter Se- *Cop. lib. i.*
midiametrum Orbis magni, id eſt, diſtantiam Solis à terra, & Semi- *Reuol. c. 6.*
diametrum firmamenti, ſiue diſtantiam ſtellæ cuiuslibet fixæ à ter- *& alibi.*
ra, nulla intercedit proportio vllo modo fenſibilis; atqui hoc abfur-
dum eſſe, & alia abſurda gignere, è ſequentibus manifeſtum fiat.
Secundum Copernicæam ſententiam [*a*] continet   *a*] *Maſlin.*
Semidiameter terræ milliaria germanica   860. *in Appédi-*
Maxima [*b*] Solis diſtantia à terra, Semidiametros terræ   1208, *ce ad Nar-*
                                                                   rationem

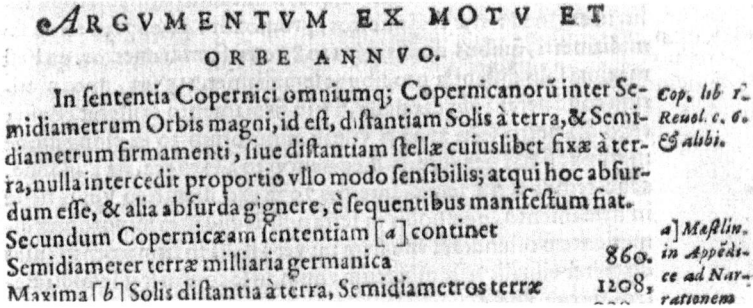

FIGURE N-5. The beginning of Locher's "Argument based on the Earth's annual motion around the sun" in his original text. Here is an example of a place where in translation I have simplified Locher's text significantly. A close translation of this (excluding marginal notes), would read:

> *In the opinion of Copernicus and of all the Copernicans, between the Semidiameter of the great Orb, that is, the distance to the Sun from earth, and the Semidiameter of the firmament, or the distance of any fixed star from earth, there exists no proportion sensible by any manner; but from the following it may be made manifest this to be absurd, and to give birth to other absurdities. Following the Copernican opinion the Semidiameter of earth contains—860 german miles. The greatest distance of the Sun from earth [contains]—1208 Semidiameters of earth, that is—1,038,880 common german miles, or one thousand thousand, thirty-eight thousand, eight hundred eighty miles: the proportion of that to 860 is the same that 1208 is to 1. But 1 to 1208 is still sensible in this our subject of consideration, from the parallax of the sun, noted by all, granted by all, and indeed by Copernicus and by the minions of him.*

Further on in this section Locher will even write out 13,133,376 as *tredicies millenae millies, centies et tricies ter millenae, centum et septuaginta sex,* or "thirteen thousand thousand, one hundred and thirty-three thousand, three hundred and seventy-six."

distance to them. Locher notes astronomers being in agreement that the distance of the sun is not sufficiently large for Earth's size to be negligible by comparison.

62. Locher uses the word *asseclis,* which means "followers" in a contemptuous sense. Copernicus notes in book 4, chapter 19 of *De Revolutionibus* how "even the sun has some parallax" (Copernicus 1995, 208).

63. I have translated this section very freely, to the point of paraphrasing the mathematics, as Locher does things like write out each number in words and apply unusual formatting. See Figure N-5.

64. This value is far too small. The diameter of the sun actually exceeds 200 semidiameters of Earth. However, this value is consistent with Locher's stated

value of 1,208 Earth semidiameters for the distance of Earth from the sun. As Locher notes, the apparent diameter of the sun is ½ degree, or 0.008727 radians. At a distance of 1,208 terrestrial semidiameters, this translates into 1,208×0.008728 = 10.54 semidiameters.

65. Note on angular measure: One *degree* is 1/360 of a revolution or of a full circle; one *minute of arc* is 1/60 of a degree; one *second of arc* is 1/60 of a minute or 1/3600 of a degree.

66. See Brahe 1915/1602, 424–31; Graney 2015, 34.

67. Note that Locher is speaking from a geocentric point of view here, so the star would extend from Earth past the moon, Mercury, Venus, the sun, and Mars, toward Jupiter.

68. Locher actually references book 12, proposition 16 of Euclid. However, this seems to be an error, since this proposition pertains to inscribing a polygon in a circle. See Euclid 1956, 3:423. These lines contain a number of errors, one of which is marked in Locher's errata section. The correct reference is certainly book 12, proposition 18, which pertains to the volumes of spheres: "Spheres are to one another in the triplicate [cubic] ratio of their respective diameters." See Euclid 1956, 3:434.

69. See Brahe 1610, 481, "De Affixarum Stellarum veris Magnitudinibus, Autoris censura."

70. See Copernicus 1995, 14. A better citation might have been the end of chapter 10, where Copernicus marvels at the size of the heliocentric universe in his system and associates this size with the amazing work of God. See Copernicus 1995, 27.

71. Rheticus's *Narratio Prima,* page 118—that is, Kepler 1596, 118—where Rheticus marvels at the vastness of the Copernican universe and the workmanship of God. See also Rosen 1959, 144–45.

72. Certain Copernicans appealed to the power of God as a solution to the problem of giant stars that the Copernican theory introduced. See Graney 2015, 76–85.

73. Translation of *The Sand Reckoner* from Heath 1897, 221–22.

74. Simple observations of the heavens show that Earth is of negligible size compared to the distance to the stars. See Figure N-6.

75. The italics are Locher's addition. Locher omits a portion of Brahe's words. The full quotation from Brahe is "Copernicus, lest he impinge into this absurdity [that Earth moves around the sun without any annual parallax being visible in the stars], and impute to the fixed stars some apparent instability as a result of the annual revolution of Earth, has preferred rather to grant another sort [of absurdity], no less inharmonious and incredible, a vastness of such size evidently to be encompassed between most distant Saturn. . . ." The rest of the quotation is

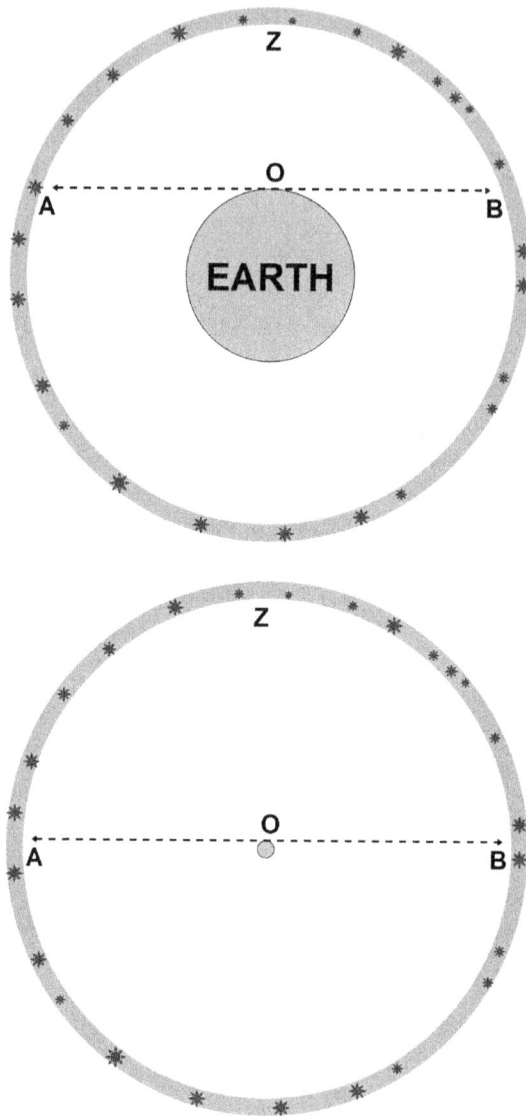

FIGURE N-6. *Top*: Were Earth of significant size compared to the distance to the stars, then an observer at O would see only the span of the starry heavens from A on one horizon, through the zenith Z, to B on the opposite horizon—well less than half the celestial sphere could be seen at any given moment. *Bottom*: That half the heavens are in fact seen at any given moment demonstrates that Earth is in fact not of significant size compared to the distance to the stars. Here the span of the starry heavens A through Z to B, as seen by the observer at O, is nearly half the celestial sphere, and were Earth so small as to be nearly a point in this figure, the AZB span would be exactly half the celestial sphere.

as Locher states. See Brahe 1610, 685. See also the introduction to this book in regard to Brahe and the problem of stellar size in the Copernican system.

76. Galileo includes in his 1632 *Dialogue* an extensive discussion of this section of Locher's *Disquisitions*. See Galilei 2001, 268–69.

77. See Kepler 1604, 309, where Kepler refers to "the line of the perpendicular, following which all heavy things are carried downward" (Kepler 2000, 320). See Stevin 1605, definitions 2 and 5 (pages 5–6); these pertain to weight and the definition of horizontal and vertical. Simon Stevin had "frankly and wholeheartedly adopted the Copernican doctrine" (Dijksterhuis 1955, 12).

78. Figure 14-1 is mischaracterized in a remarkable fashion by both Galileo and Stillman Drake, who translated Galileo's *Dialogue* into English. Regarding the figure, Galileo writes, "My, what pretty pictures; what birds, what balls! And what are these other beautiful things? . . . balls which are coming from the moon's orbit . . . a snail which they call *buovoli* here in Venice; it is also coming from the moon. . . . Oh, indeed. So that is why the moon has such a great influence over these shellfish" (Galilei 2001, 269). Drake adds a note regarding the *buovoli*, commenting on "the amusing recital here of the many irrelevancies in the engraved plates" of Locher's book. See Galilei 2001, 562. In fact, the snail is not dropped from the moon's orbit, but by a bird. Nor are the birds and balls irrelevancies; Locher will use them all to make his point about common motion on a rotating Earth.

79. In modern terms these are 25,000 miles in one day, 1,042 miles in one hour, 17.4 miles in one minute, and 0.29 miles in one second. Locher attributes the 4,166 figure to Oronitus.

80. In modern terms these are 16,728 miles in one day, 697 miles in one hour, 11.6 miles in one minute, and 0.19 miles in one second.

81. Locher is assuming for the sake of argument that the rate at which a ball drops to Earth is the same as the rate at which it would circle Earth in the Copernican opinion. He envisions the balls as "in concavo sphaerae Lunaris positas," that is, "stationed in the concave of the sphere of the moon" (under the idea that the moon is carried on a sphere) and thus as being both at the distance of the moon and sharing the velocity of that sphere, which would be the moon's orbital velocity. The moon orbits Earth once every 27.3 days (measured with respect to the stars—this is shorter than the time for the moon to complete a cycle of phases, which is determined by the moon's motion with respect to the sun). Thus, 27.3 days would be required for the moon to circle from A through η and back to A in Figure 14-1. It follows that the moon would complete a quarter of this cycle—it would circle from A to C (behind the center of Earth in the figure)—in 27.3/4 = 6.825 days. Equating the linear motion from A to C to the circular motion from A to behind C, the time of fall of a ball from A to C is also 6.825 days. Locher tends to

drop fractional amounts, and A to B is not quite as far as A to C, so he arrives at six days. Locher's insight turns out to be quite good: the modern understanding of an orbit is that indeed it is essentially a continual fall (and Locher will connect a Copernican orbit to a fall in disquisition 15), and six days is not an unreasonable value for the time required for an object to fall from the height of the moon. In the *Dialogue*, Galileo criticizes the six-day value, having calculated the time to fall from the height of the moon as being under four hours. See Galilei 2001, 256, 262. Galileo's time of fall is much too short. Note that other modern writers have interpreted the six-day value differently—see Finocchiaro 2014, 141; Heilbron 2010, 436n93. This section is difficult, in part because Locher uses a parenthetical phrase that lacks a closing parenthesis.

82. *De Revolutionibus*, book 4, chapter 17: "Distance of the Moon from the Earth, etc." See Copernicus 1995, 205–7. There Copernicus calculates that the distance to the moon is 56 42/60 (56.7) times the semidiameter of Earth—a value close to that produced through modern measurements. Using the circumference of 5,400 German miles for the circumference of Earth given previously, the semidiameter of Earth is $5,400/2\pi = 859.44$ German miles. The distance to the moon is then $859.44 \times 56.7 = 48,730$ German miles, and the circumference of the moon's orbit is $48,730 \times 2\pi = 306,180$ German miles. As ball A supposedly travels around Earth once in twenty-four hours, in one hour it travels $306,180/24 = 12,758$ German miles, close to the value Locher gives of 12,600. Maestlin's appendix to the *Narratio Prima*—see Kepler 1596, 167—is also cited, which contains a discussion on distances and the moon. I have not been able to determine how the exact 12,600-mile value was obtained.

83. Circling around Earth at the same rate that Earth turns, so that it stays above the same spot on Earth.

84. Locher's argument is not trivial. Experience seems to indicate that heavy objects fall straight down, whereas light materials such as smoke rise straight up. On a rotating Earth, falling and rising are complex motions: the spiral path of the ball falling for six days is an extreme case, but the point is valid. Galileo argued in the *Dialogue* that Locher's falling balls would not remain over the same spots as they fell, but would rather run ahead of the rotating Earth. See Galilei 2001, 271. The implication of this, however, is that to some extent falling bodies *do not* move straight down. Rather, they must be deflected as they fall (and their motion remains fairly complex). The apparent lack of any such deflection was cited in 1651 by Giovanni Battista Riccioli as evidence for Earth's immobility, and in 1679 Isaac Newton proposed to Robert Hooke that the deflection of a falling body might serve to prove Earth's rotation. This deflection does in fact exist and is part of what is known as the Coriolis effect (see Figure I-4). However, it eluded detection until the nineteenth century. See Graney 2015, 118–28.

85. Galileo devotes considerable space in his *Dialogue* to addressing these questions. See Galilei 2001, 274–87. Locher uses an awkward numbering system for these questions that does not group them in a manner consistent with the text of the questions. Here I maintain the order of the questions but opt for a bulleted list system to group the questions in accordance with the flow of Locher's words.

86. Locher seems to be referring to the suggestions by Copernicus that "perhaps the movement of the air is an acquired one, in which it participates without resistance on account of the contiguity and perpetual rotation of the Earth," and that "the highest region of air follows the celestial movement." See Copernicus 1995, 17.

87. The surface of Earth moves at differing speeds—fastest at the equator, zero at the poles—and the air follows that surface everywhere.

88. Points higher in the air move faster than points closer to Earth's surface. As Earth rotates, a higher point traces out a larger circle, and moves with greater speed, than a lower point.

89. Locher uses alliteration and a little crudeness here—"plumam et plumbum? fumum et fimum?"—and so I have done the same.

90. See Copernicus 1995, 17.

91. Note that in the physics of Locher's time, the four elements (earth, water, air, and fire) from which physical objects were comprised had entirely different properties. Thus the element earth had weight, or heaviness, or *gravity*, a natural tendency to move downward. By contrast, the element fire had lightness, or *levity*, a natural tendency to move upward. Gravity would be apparent in a stone; levity would be apparent in smoke. Whereas in modern physics all objects have some properties in common, such as mass, in Locher's time this was not understood to be the case.

92. This is the first of three questions that pertain to the fact that points on and near a rotating Earth do not share a completely common motion. In the case Locher mentions here, if one object is above another, then the one which is higher is farther from the center of Earth. Therefore, as Earth rotates, the upper object traces out a larger circle and moves with greater speed than the lower object. In the next two questions, Locher raises the point that objects near the equator move at different speeds, and in different manners, than objects near the pole. Objections that a rotating Earth produces different kinds and speeds of motion at different places on Earth, and that this leads to difficulties for the Copernican hypothesis, date at least to Tycho Brahe. Such objections in essence anticipate the Coriolis effect (see Figure I-4) caused by Earth's rotation. See Graney 2015, 37–43, 115–28.

93. Locher supposes that if Earth rotates then the rocket should be seen to curve, but he is not clear as to why. He may think this is on account of the rocket's changing distance from the center of Earth and its passing through atmosphere that is moving faster and faster (as he sees it), or on account of what is now called

the Coriolis effect (see Figure I-4). This would be consistent with the other questions he raises here. But his description is too muddled to be sure.

94. Locher provides the reference to book 1, chapter 8, and only the few words in italics, which on their own are not that illuminating. However, from the opinions Locher attributes to Copernicus's *asseclae* (sycophants, minions), it seems Locher expects the reader to be familiar with this chapter. Thus I have added some additional material from that part of chapter 8, indicated with brackets, so as to directly convey the information Locher is implicitly conveying. See Copernicus 1995, 18.

95. Locher is approximating $\pi = 3$.

96. 6,233,280 divided by 365 is actually 17,077.

97. Macrobius 1556, 107–8. Translation from Macrobius 1952, 182.

98. See Clavius 1611, 17–72, which discusses the composition and location of the globe of the earth.

99. See Stevin 1605, 6. This is Stevin's definition of center of gravity as the one point in a solid from which, were the solid to be suspended from it, the solid would remain at rest regardless of orientation. See Dijksterhuis 1955, 38.

100. There are two remarkable features to Locher's discussion here. First, Locher proposes that the circular orbital motion of a massive body might be explained in terms of a centripetal (center-seeking) force that drives a perpetual falling action. Locher's explanation is remarkable in that it has much in common with the Newtonian explanation of orbits, which also treats an orbit as a perpetual fall under a central force. Second, Locher leads the reader to this explanation by envisioning a limiting case. In Figure 15-1 he asks the reader to envision the surface of Earth being smaller and smaller, being first curve MN, then OP, then QR. In each case the ball falls in a curve. The limiting case is that Earth is vanishingly small—and Locher supposes the falling process to still occur in this case.

101. See Clavius 1611, 17–72, but especially 19, where he speaks of the difference in the heaviness of earth versus water, and 20 and 46, where he provides diagrams of the order and position of the elements.

102. The value Locher gives here is actually 760, which conflicts with the earlier statement (see disquisition 14) that the circumference of Earth is 5,400 German miles. A circumference of 5,400 German miles would imply a semidiameter of 860 German miles. Locher's reference is probably to Clavius 1611, 62, which discusses how Earth is a globe for astronomical purposes, even if it is not a geometrically perfect globe. However, I have not been able to find any explicit mention within Clavius of the numbers Locher provides.

103. Book 3, proposition 7: "If on the diameter of a circle a point be taken which is not the center of the circle, and from the point straight lines fall upon the circle, that will be greatest on which the center is, the remainder of the same di-

ameter will be least, and of the rest the nearer to the straight line through the center is always greater than the more remote. . . ." See Euclid 1956, 2:14.

104. All astronomers would indeed have known that the sun is closer to Earth. Earth's annual orbit around the sun is slightly elliptical, which means that Earth is closer to the sun at some times of the year than at others. The time during which Earth is closest to the sun is during the northern hemisphere's winter. For this reason the sun seen from Earth appears larger during the northern hemisphere's winter, and the apparent size difference can be measured telescopically.

105. The moon at quarter sits at the corner of a right angle formed by Earth, the moon, and the sun. The hypotenuse of this very long, skinny triangle is the distance from Earth to the sun and is thus longer than the distance from the moon to the sun. So the quarter moon is slightly closer to the sun than is Earth.

106. "Book 1, proposition 35" is all the citation Locher provides. However, a discussion by Kepler of the subject of light from planets and the moon that matches what Locher describes here can be found in Brackinridge and Rossi 1979, 94–95.

107. As mentioned previously, the sun is more distant during the northern hemisphere's summer, closer in its winter, so between June and December the sun approaches Earth (as seen by an observer on Earth), and between December and June the sun recedes from Earth.

108. As with the gnomon (disquisition 15), Locher is here again asking the reader to envision a limiting case.

109. Locher is apparently of the opinion that light could not travel absent a medium, an opinion that would persist among physicists up through the time of Einstein.

110. Locher includes two abbreviated citations of Euclid book 1, with numbers 1 and 9, but I cannot determine to what in Euclid he is trying to refer.

111. Locher describes these stars as "valde immersas esse"—deeply immersed in the firmament.

112. Homocentric: one single center of motion, Earth. Aristotle described a homocentric system developed by Eudoxus. See Aristotle 1984, 2:1,696–97. Fracastoro proposed a system that hearkened back to Aristotle. The 1574 work to which Locher is referring is probably a certain posthumous collection of Fracastoro's works—see Fracastoro 1574, 1–48. Fracastoro originally published his theory several decades earlier, just prior to Copernicus's 1543 De Revolutionibus.

113. Eccentrics: off-center circles or spheres. Epicycles: secondary circles that turn on larger primary circles, called deferents, used to explain the motions of the planets among the stars. See Figure I-3 for illustrations of an eccentric and an epicycle.

114. For example, the orbit of Jupiter's moon Callisto can be thought of as a smaller secondary circle (epicycle) riding on the larger primary circle (deferent) that is Jupiter's orbit. Note that while the telescope is commonly viewed as undermining Ptolemaic ideas, here Locher sees the telescope as providing observational confirmation of the existence of epicycles. See the introduction for further discussion.

115. The apparent size of both the moon and the sun do change over time, as the distance between Earth and these bodies is not constant, owing to the elliptical orbit of Earth about the sun and of the moon about Earth. Fracastoro attributed these changes to refraction by certain celestial orbs.

116. Simple diagrams of the Copernican system show all the planets going around the sun in concentric circles, but Copernicus used both eccentric circles and small epicycles in his system. These in essence approximated the actual elliptical paths of the planets around the sun.

117. King Alphonso X of Castile, noted for his support of astronomy and the production of the Alfonsine Tables (1252).

118. Peucer and Dasypodius 1568.

119. Leopold of Austria 1489.

120. Clavius 1611, 36. See also 308–13 (tables showing number of orbs). For further discussion, see also Lattis 1994, 108–17, 242.

121. The angular separation of Venus from the sun in the sky.

122. Translation from Lattis 1994, 198, with italics according to Locher. See Clavius 1611, 75. Locher omits a piece of text from Clavius at the end of the first paragraph. Thus Clavius wrote the last sentence of the first paragraph as "Consult the reliable little book by Galileo Galilei, printed at Venice in 1610 and called Sidereal Messenger, which describes various observations of the stars first made by him."

123. "Clauius suum systema abrogat."

124. Translation is taken from Stahl and Capella 1977, 333, with minor changes and italics according to Locher. See 338. Locher also has a marginal note regarding some typographical errors: "intra" for "infra"; a missing "qua"; "castiore" for "laxiore".

125. Stahl and Capella 1977, 342; Capella 1539, 350.

126. Stahl and Capella 1977, 332; Capella 1539, 337.

127. Stahl and Capella 1977, 343; Capella 1539, 350.

128. Stahl and Capella 1977, 343; Capella 1539, 351.

129. Stahl and Capella 1977, 344; Capella 1539, 352; italics added by Locher.

130. In the Ptolemaic system, each planet had its own epicycle and deferent, and the movements of the epicycles were slaved to the movement of the sun. In the Capellan/Brahean system, all the planets share the same deferent, namely the sun's orbit about Earth, thus reducing the number of circles. Furthermore, since the planets all circle the sun, the connection between the sun's motion and planetary motions is clear.

131. Locher cites this as being from Vitruvius, book 9, chapter 4. Jean Baptiste Morin also cites this phrase the same way in a discussion of different world systems—see Morin 1631, 9. But according to Vitruvius 1914, 259, this phrase is from book 9, chapter 1, paragraph 5.

132. This probably means there are fewer compared to a Ptolemaic system, where every planet has its own epicycle and deferent.

133. Number depends on how many planets circle the sun. Material is perhaps uncertain because in the Tychonic system Mars's orbit around the sun crosses the sun's orbit around Earth, calling into question whether there can be solid celestial orbs that govern celestial motions. Thickness is probably uncertain because whereas the complicated machinery of the Ptolemaic system forced certain separations between planets, this is not the case in the Tychonic system.

134. Astronomers of the time explained this through the concept of "adventitious rays" that supposedly made bright objects appear larger. See Graney 2015, 53–61. Locher will explore this more in disquisition 28.

135. The Al-Batani reference is probably Albategnius 1537, chapter 30, and in that perhaps page 40, on how the sun illuminates the moon. For the Vitellius reference Locher actually has book 4, *page* 77, but paragraph 77 seems more likely as it contains some discussion of the sun being a source of illumination, while page 77 does not contain related material—see Risner 1572, 151–52.

136. In the main text Locher uses the word *planetarum*, or "of the planets," but in the marginal note next to the text he uses the word *astrorum*, or "of the stars."

137. A telescope whose object lens has a longer focal length (and thus a longer tube) produces larger images, and so this may be what Locher writes of here. This and the following paragraphs discuss the appearance of stars seen through a telescope whose eye lens has been removed and note the effects of lens or eye problems on that appearance. Such an arrangement causes the appearance of stars to be distorted, but it renders the colors of stars easier to see.

138. The heading "Concerning the Types of Stars (or Heavenly Bodies)" introduces the subject of the remaining disquisitions. Locher placed it after the heading "Disquisition 25," but to clarify the book's structure it has been moved to precede the disquisition number.

139. The face of the moon does appear to rock slightly back and forth as seen from Earth (as though the "man in the moon" was very slowly shaking his head). Note that the moon is circling Earth on its deferent BCD in Figure 25-1. Locher's discussion is focused on the epicycle, but there is a motion on the deferent at the same time.

140. The lunar spots visible from Earth, which comprise the "man in the moon," for example.

141. See Kepler 1604, 227. Translation is from Kepler 2000, 241–42.

142. Thales and Anaxagoras were ancient writers who identified the moon

as being illuminated by the sun. Locher actually references book 9, chapter 4 of Vitruvius, but since that is on constellations and chapter 2 is on the moon, that is probably a typographical error—see Vitruvius 1914, 262. For Macrobius, see Macrobius 1556, 88–89, or Macrobius 1952, 164–65, which discusses the moon and states that the moon "could have not light except from the sun above it."

143. A geometrically perfect sphere illuminated by the sun would show one strong glint of light, in the manner of the reflective glass balls used as yard or Christmas tree ornaments. Therefore, since the visible surface of the moon is uniformly illuminated (approximately), the moon cannot be a geometrically perfect sphere.

144. That is, spherical, but not a geometrically perfect sphere.

145. That is, craters.

146. The ring effect that Locher describes here is seen even by modern amateur astronomers who observe the moon visually with their telescopes, sketch it, and describe it (see Figure N-7). Galileo, in his *Dialogue,* acknowledges that the ring is seen but states that it is an illusion: it "originates deceptively" (Galilei 2001, 107). Galileo also refers to Locher's discussion on the ring as "a lie that borders upon rashness" (Galilei 2001, 108).

147. This section is discussed in some detail by Galileo in his *Dialogue.* See Galilei 2001, 105–10. Galileo's discussion is not at all favorable, speaking through the character of Salviati of showing "the futility of the arguments of [Locher], in which there are falsehoods and fallacies and contradictions" (Galilei 2001, 107), such as the "lie that borders on rashness" comment mentioned in the previous note. Locher's discussion of the secondary light of the moon is directly opposed to Galileo's claim, introduced in his *Sidereus Nuncius* of 1610, that the secondary light is caused by the illumination of the moon by light from Earth. Galileo argued in the *Nuncius* that the secondary light weakens as the moon waxes from a crescent toward first quarter. See Carlos 1880, 30–38.

148. This idea appears in a folk name for the secondary light of the moon, especially when the moon is a young crescent: "the old moon in the new moon's arms." See Mayall et al. 1977, 55.

149. Venus shines by reflected light, so when Venus is between the sun and Earth, its illuminated face is directed toward the sun and away from Earth (and the moon).

150. Whether it be a more distant "wandering star" (such as Jupiter) or a still more distant fixed star (such as Arcturus), most of the brighter "stars" fall more or less into conjunction with the sun at one time or another and thus are subject to this argument to at least some extent.

151. See Macrobius 1556, 88–89, or Macrobius 1952, 164–65, which specifies that the moon "could have no light except from the sun" and allows light "to penetrate to such a degree that it sends it forth again."

FIGURE N-7. A modern sketch showing the brighter ring along the dark limb of the moon, described by Locher and illustrated in Figure 26-1. The artist is Deirdre Kelleghan, an accomplished visual astronomer who is a coauthor of a book on sketching the moon (Handy et al. 2011) and blogs for the Vatican Observatory. Regarding this sketch, Kelleghan notes that "this brightness on the lunar limb in my pencil drawing . . . [is] part of my moon drawing habit because it is exactly what I see. What is going on to manifest this limb glow is not clear. . . ." Kelleghan notes that the glow can be very obvious. It also appears in the work of others who draw the moon. See Kelleghan and Graney 2016.

152. Locher cites book 3 of Cardano's *De Subtilitate*—see Cardano 1554, 105.

153. Here Locher argues entirely from a traditional Aristotelian viewpoint: Earth cannot possibly reflect the sun's light as well as the moon reflects it; celestial bodies are made of a different sort of stuff than Earth. Locher does not use such arguments often, and he will presently return to arguments based on observations and measurements and mathematics, but in this case he sounds much like Galileo's Simplicio—see, for example, Galilei 2001, 81, where the character Simplicio argues that Earth is unfit to reflect sunlight while the moon is quite fit to do so.

154. When the sun is just above the horizon, such as at dawn or dusk, its rays strike Earth at an oblique angle and thus illuminate the surface of Earth less than when the sun is high in the sky. Thus the supposed rays of Earth should likewise illuminate less those parts of the moon where Earth would be seen as just above the lunar horizon. At first glance this seems like an effective argument, but this effect is not seen in the full moon, when it would apply to the rays of the sun illuminating the moon. In fact, while the rays of the sun do illuminate the surface of the full moon less at its edges, for the reason Locher states, we also see the surface of the full moon at the edges through an oblique angle and thus to a certain extent compressed (this effect is called foreshortening, and it causes a round lunar feature seen near the edge of the moon to appear oval). The reduced surface illumination and apparent surface compression effectively cancel each other out, with the result that the full moon appears uniformly illuminated as seen from Earth.

155. This sentence appears to be out of place and might be more suitable a few paragraphs earlier.

156. The Vitellius reference seems to be paragraph 77 of "Vitellonis filii Thuringorum et Polonorum opticae liber quartus"—see pages 151–52 in Risner 1572—which relates the appearance of solar eclipses to the phenomenon of light passing through an obstructing translucent sphere. There are effects in total eclipses of the sun that produce a ringlike appearance, such as the "diamond ring effect" shown in Figure N-8, which Locher may be interpreting as being further manifestations of the illuminated ring illustrated in Figure 26-1. The Reinhold reference is probably to Reinhold's "De Illuminatione Lunae," which includes a discussion of the moon during a solar eclipse as evidence of the moon's illumination—see Reinhold and Peurbach 1580, 160.

157. For a different, looser translation of this paragraph, see Galilei 2001, 105.

158. I have not been able to determine why Locher cites the astronomical *Phenomena* by Aratus, which contains no clear reference to the secondary light of the moon—see Aratus 1898. He gives no specifics regarding the Cleomedes reference, but Cleomedes does discuss this—see Cleomedes, Bowen, and Todd 2004, 142. The parts he cites from Vitellius (see Risner 1572, 151–52), and Aquilon (see Aquilonius 1613, 419) seem to treat the manner by which light might pass through

FIGURE N-8. Modern photograph of the "diamond ring effect" in a total eclipse of the sun. Note some similarity to the ring of illumination illustrated by Locher in Figure 26-1. Image courtesy of Wikimedia Commons.

a translucent sphere and to discuss the moon specifically, although I have not made a translation of these sources.

159. See Macrobius 1556, 89. Translation from Macrobius 1952, 165.

160. According to the calculations of the Stellarium computer software package, an eclipse was seen in Ingolstadt not on 29 May, 1612, but on 30 May. However, it was only a partial eclipse of the sun, so the diamond ring effect (see Figure N-8) was not visible. I have seen a number of partial solar eclipses and am not familiar with any partial eclipse effect such as Locher describes here as being widely visible. However, it is clear that Locher believes eclipse observations to add support to the existence of an illuminated lunar ring such as is illustrated in Figure 26-1.

161. Locher's words here are "tanquam Rex medio planetarum quinque consistens loco"—"just as a king, standing at the middle point of the five planets."

While earlier Locher did not absolutely endorse the Tychonic system (see the last paragraph of disquisition 22), this statement is Tychonic (or, as Locher might prefer, Capellan) in its view.

162. The nova of 1572, which was bright enough to see during the daylight. Today astronomers understand that nova to have been an exploding star.

163. Here Locher refers to discussions regarding telescopic observations of the sun, written by his mentor, Christoph Scheiner, under the pseudonym "Apelles," starting in November of 1611, which were answered by Galileo. See Van Helden and Reeves 2010.

164. Solar observation by telescope was accomplished by means of using the telescope to project an image of the sun onto a screen.

165. The words Locher uses are *maculae* and *faculae*. *Maculae* translates as "spots" or "blemishes." Astronomers today regularly refer to the sun's "spots." *Faculae* translates as "little torches," or "splinters used as a torches"—almost like igniting matchstick heads. Astronomers still use the term *faculae* to refer to bright patches on the surface of the sun. I have opted, for the sake of clarity and brevity, to render the term "blazes" with the intent of conveying an image of small, more brightly glowing features on the sun.

166 . Here Locher means that the positions of the spots do not change systematically over a day, as they would on account of the change in the relative positions of the spots, the sun, and Earthbound observers that would result from the rotation of Earth (or the circling of the sun, depending on the world system to which one subscribes), were the spots bodies standing a good distance in front of the sun. Sunspots do change position during a day owing to the rotation of the sun and to various motions caused by the fluid nature of the solar surface. Scheiner had discussed this previously—see Van Helden and Reeves 2010, 68.

167. Galileo's 1613 *Istoria e dimostrazioni intorno alle macchie solari e loro accidenti,* in which he discussed sunspots in answer to Apelles (Christoph Scheiner). See Van Helden and Reeves 2010, 246–48, 408.

168. See Van Helden and Reeves 2010, 307 for a slightly different translation of this entire paragraph; also see Blackwell 2006, 70.

169. See Van Helden and Reeves 2010, 308 for a slightly different translation of this entire paragraph.

170. Translation of Galileo quotation taken from Van Helden and Reeves 2010, 281, with italics added to agree with Locher.

171. Locher does not explicitly attribute this refraction to the air, but indeed it does exist and is caused by the atmosphere. Locher's discussion is brief but correct: the same vertical distortion that squashes the sun into an elliptical shape at the horizon indeed does cause it to appear higher in the sky than it would in the absence of an atmosphere, and to appear above the horizon for longer (and thus

cause the days to be longer) than it would in the absence of an atmosphere; it also distorts the positions of stars, and, in causing them to appear above the horizon for longer, causes them to appear to linger near the horizon. This distortion is a squashing of the sun "from below": the lower edge of the sun's circumference is refracted "up"—above the horizon; the upper edge of the sun's circumference is not refracted "down" toward the horizon. Actually, all parts are refracted "up"—the squashing effect comes from the fact that the effect is stronger closer to the horizon, so the lower edge of the sun is refracted "up" more than is the upper edge.

172. This phenomenon is nicely illustrated in the opening moments of the 1994 Disney movie *The Lion King,* which features a sun just beginning to rise.

173. Locher is probably referring here to observing these tower globes by means of a telescope.

174. The brightening is the result of a roughly round blob of refractive material concentrating light like a lens; the strangling of the light would be from dispersing or redirecting the ray of light away from the eye. Locher's discussion generally agrees with modern ideas about atmospheric "seeing."

175. Here Locher is probably referring to shadows cast through fumes rising from a stove and how they will fluctuate at their edges.

176. Locher cites page 44 of the second booklet by Apelles—Scheiner 1612, 44—but this reference seems to be in error, for there is no mention of scintillation or anything similar at this location. I could find no mention of scintillation in the sun and stars within this work, even searching electronically both the original work and a translation—see Van Helden and Reeves 2010, 183–230, especially 222–23, where is found the translation of Scheiner 1612, 44.

177. All these quotations are from Risner 1572, 449.

178. Locher rejects various explanations for scintillation. Various secondary sources note that astronomers including Brahe, Kepler, and Galileo, among others, attributed scintillation to the rotations, spasms, vibrations, etc. of the stars. See, for example, Monaco 1990, 819–20.

179. Venus is far brighter than Sirius, and indeed it is often said to be bright enough to cast shadows, but I can find no writers who suggest that Venus may be bright enough to read by.

180. In this section Locher assumes his reader to be familiar with Simon Marius's 1614 *Mundus Jovialis (The World of Jupiter),* and in particular with a few pages of its preface—see Marius 1614, fifth through ninth pages of "Praefatio ad Candidum Lectorem." Here he is probably making reference to Marius's complaint about those who will criticize him and accuse him of gross errors—"Praefatio," sixth page.

181. See Risner 1572, 449.

182. See Risner 1572, 449.

183. Marius writes of seeing the colors, green, gold, red, and blue when observing Sirius by means of a telescope with no eye glass. See Marius 1614, "Praefatio," sixth page. For secondary sources that mention Marius in this matter, see Anonymous Reviewer 1853, 197, or Knight 1868, 458.

184. See Marius 1614, "Praefatio," seventh page.

185. In an effort to render disquisition 36 readable by as broad an audience as possible, I have inserted explanatory phrases not in Locher's Latin text. Here I identify that the "object" of the optical system Locher describes (and illustrates in Figure 36-1) is the sun, while the cloud example is my addition to clarify what Locher is discussing regarding division of the cones.

186. Were a cloud to pass over half of the sun, this would be seen in all three images.

187. The comment about left and lower semidiameters I have added in an attempt to provide clarity.

188. See Clavius 1611, 42. For a slightly different translation, and some discussion on Clavius and the order of the planets, see Westman 2011, 209. The other sources cited in this quotation do not specify this order exactly. Plato specifies the order as "the moon in orbit nearest the earth; then . . . the morning star and the star said to be sacred to Hermes" (Plato 1961, 1,167, *Timaeus* 38c). *On the Heavens,* book 2, chapter 10 contains only a general discussion that the order follows speed, while *Meteorology,* book 1, chapter 4 does not contain a clear reference to motions of these bodies. See Aristotle 1984, 1:480 and 559–60. *On the World,* which is probably not from Aristotle, gives the order stated—see Aristotle 1931, chapter 2, 392a20–30.

189. See Aristotle, Apuleus, and Budaeus 1591, 32–40. This is a version of *On the World* with notes and comments.

190. Venus and Mercury circling the sun (while the sun circles Earth, as Locher would see it) results in a much smaller machinery of the heavens than if Venus and Mercury each have their own orbit around Earth, plus their own epicycle, in addition to the sun having its own orbit around Earth.

191. Simon Marius, author of *The World of Jupiter,* also published in 1614. See Marius and Prickard 1916, 372–73.

192. This observation, which Locher dates as 9:15 p.m. April 29, must have been made just before that moon (Europa) emerged from Jupiter's shadow. According to the Stellarium planetarium software package, Europa began to emerge from the shadow at about 9:10 p.m. local time and was fully clear of the shadow at about 9:15 p.m. These times are modern measurements, given by Stellarium for Ingolstadt's time zone, but Stellarium also shows the sun in Ingolstad to have been almost on the meridian at 12:00 p.m., so modern time and time as measured by the sun should match to within a few minutes. Compare Locher's illustrations (Figure 39-1) to Stellarium (Figure N-9). Moreover, note that while Stellarium

FIGURE N-9. Stellarium simulation of the appearance of Jupiter and its moons at the times shown by Locher for the night of April 29–30, 1614 (see Figure 39-1).

shows all four moons to have approximately the same appearance, Locher shows size differences from moon to moon. In fact, Locher's illustrations are more accurate than Stellarium's graphics regarding the appearance of the moons. For example, Locher's 11:00 p.m. sketch shows the leftmost and lower moon (Callisto) to be smaller than the one above and slightly to the right of it (Ganymede), which is about the same size as the next moon to the right (Io), with the moon closest to Jupiter (Europa) smaller again. While Stellarium shows these as all about the same size, it calculates their magnitudes as follows: Callisto, 6.60; Ganymede, 5.59; Io, 5.98; Europa, 6.26. As a higher magnitude means less bright—or as Locher would see it, less large—Locher and Stellarium are in agreement.

193. Galileo's discourse on floating bodies—see Galilei 1663, 1.

194. This may refer to Marius's mention in the preface that Galileo's observations "were of great assistance to me" (Marius and Prickard 1916, 373).

195. Locher does use dismissive language to refer to Marius. He says Marius "sponged" off Galileo—the verb he uses is *hausit*, meaning to draw off, drain, borrow, swallow up, consume, devour. He refers to measurements according to Galileo and then according to *alterum*—"the other." He also refers to the measurements "Galilaei aemulo"—"by the emulator of Galilei."

196. See Marius and Prickard 1916, 407–8. Marius describes using long periods of time to measure the periods of revolution of Jupiter's satellites. He also says he used observations of the satellites when they were near Jupiter, probably much like Locher describes.

197. For the fast-moving inner moon B, this would be hours; for the slow-moving outer moon E, this could exceed a day.

198. As noted in the introduction, here Locher is proposing a program of further research for the astronomical community ("all astronomers"). Catching a moon coming out of eclipse requires patient observation.

199. See Van Helden and Reeves 2010, 297–99.

200. There is a certain wordplay here that I have lost in my effort to produce a translation that is accessible to a broad readership. Jupiter is high above us in the sky, and therefore the far side of Jupiter is higher above us than the near side. Likewise, the north side of the Jovian system is the "above" side in Figure 39-1, and the south side is the "below" side. Thus, what Locher really writes, in a manner befitting the classic Fibber McGhee and Molly routine about the lower price of an upper berth on a train (Fibber McGhee and Molly 1949), is that when a Jovian satellite is lower, it is higher; and when it is higher, it is lower (when on the south side, it is on the far side; and when on the north side, it is on the near side).

201. Jupiter does not exactly follow the ecliptic (the path traced by the sun) through the constellations of the zodiac, but rather deviates from it slightly, traveling to the north of it at some times, to the south at others. The line $\theta\iota$ as seen in Figure 40-1a, and the horizontal lines through Jupiter in the diagrams in Figure 39-1, are roughly parallel to the ecliptic.

202. According to calculations using the Stellarium planetarium software package, Jupiter was close to the ecliptic and close to Earth (undergoing retrograde motion and well-placed for observation) in the summers of both 1616 and 1617—to the north of the ecliptic in 1616, to the south in 1617, crossing the ecliptic on October 19, 1616.

203. Given Locher's solid analysis of the Jovian system as seen through a telescope, it is remarkable that he here uses Tycho Brahe's value for the diameter of Jupiter, which was made prior to the advent of the telescope. The diameter of Jupiter does not exceed one minute as measured telescopically. Simon Marius reports this correctly in his *World of Jupiter*—see Marius and Prickard 1916, 374. See also Van Helden 1985, 71.

204. See Marius and Prickard 1916, 411. Galileo mentions this idea in his *Starry Messenger*—see Galilei 1957, 57.

205. Locher is referring to Simon Marius, who advocated this in *The World of Jupiter*, writing that "these Jovian wanderers are illuminated in two ways, both by the Sun and by Jupiter. But the power of Jupiter to throw out his borrowed light to his satellites is very feeble" (Marius and Prickard 1916, 443).

206. A satellite that is on the far side of Jupiter will receive little light from Jupiter, as the far side of Jupiter is dark. A satellite that is on the near side will receive light from Jupiter's illuminated face, but the side of the satellite that receives that light is the side that faces away from Earth.

207. This would occur when all four satellites are simultaneously at one of their stationary points, that is, at their points of greatest separation from Jupiter as seen from Earth.

208. This would occur when all four satellites are simultaneously at the same distance to the east or west of Jupiter. The inclined plane of the orbits of the satellites would cause them to be arranged along a north–south line.

209. The text here actually reads "fourteen", but given the period provided in disquisition 40, I am assuming that to be a typographical error.

210. See Van Helden and Reeves 2010, 295–96.

211. As noted in the introduction, here Locher is proposing a program of further research. See Figure I-3.

212. The "Great Orb"—the Copernican motion of Earth around the sun. See disquisition 13.

213. Locher uses the word *Telioscopio* here—the only use in the book.

214. According to the Stellarium software package, the positions of the moon, Venus, and Saturn were much as Locher's figure indicates following sunset on February 13, 1614. The phases of Venus and the moon were also as he indicates.

215. Duoay-Rheims translation, which is of Locher's time.

WORKS CITED

Albategnius. 1537. "De scientia stellarum." In *De motu stellarum*, translated by Plato of Tivoli. Nuremburg.

Anonymous Reviewer. 1853. "'On Twinkling' by M. Arago." *Monthly Notices of the Royal Astronomical Society* 13(6): 197–201.

Aquilonius, Franciscus. 1613. *Opticorum libri sex*. Antwerp.

Aratus. 1898. *The Phenomena and Diosemeia of Aratus*. Translated by John Lamb. London: John W. Parker.

Aristotle. 1931. *The Works of Aristotle, Translated into English under the Editorship of W.D. Ross*. Translated by E. S. Forester. Vol. 3. Oxford: Clarendon Press.

———. 1984. *The Complete Works of Aristotle*. Edited by Jonathan Barnes. 2 vols. Princeton: Princeton University Press.

Aristotle, Apuleus, and G. Budaeus. 1591. *De mundo, graece: Cum duplici interpretatione Latina*. Leyden.

Blackwell, Richard J. 2006. *Behind the Scenes at Galileo's Trial*. Notre Dame, IN: University of Notre Dame Press.

Brackinridge, J. Bruce, and Mary Ann Rossi. 1979. "Johannes Kepler's 'On the More Certain Fundamentals of Astrology, Prague 1601.'" *Proceedings of the American Philosophical Society* 123: 85–116.

Brahe, Tycho. 1610. *Astronomiae instauratae progymnasmata*. Uraniburg.

———. 1915/1602. *Astronomiae Instauratae Progymnasmata, Pars Secunda*. Vol. 2 in *Tychonis Brahe Dani Opera Omnia*, edited by J. L. E. Dreyer. Copenhagen.

Capella, Martianus. 1539. *De nuptijs Philologiae, libri novem*. Lugdunum.

Cardano, Girolamo. 1554. *De subtilitate libri XXI*. Basel.

Carlos, Edward Stafford. 1880. *The Sidereal Messenger of Galileo Galilei*. Translated by Edward Stafford Carlos. London: Rivingtons.

Cicero. 1884. *De natura deorum*. Translated by H. Owgan. London: James Cornish & Sons.

Clavius, Christoph. 1611. *Opera Mathematica: V tomis distributa, etc.* Vol. 3, *Commentarium in Sphaeram Ioannis de Sacro Bosco & Astrolabium*. Mainz.

Cleomedes, Alan C. Bowen, and Robert B. Todd. 2004. *Cleomedes' Lectures on Astronomy: A Translation of the Heavens*. Berkeley: University of California Press.

Copernicus, Nicolaus. 1995. *On the Revolutions of Heavenly Spheres*. Translated by Charles Glenn Wallis. Amherst, NY: Prometheus Books.

Danielson, Dennis. 2014. *Paradise Lost and the Cosmological Revolution*. New York: Cambridge University Press.

Dechales, Claude Francis Milliet. 1690. *Cursus seu mundus mathematicus*. Vol. 4, *Complectens musicam, pyrotechniam, astrolabium, gnomonicam, astronomiam, astrologiam, tractatum de meteoris, & kalendarium*. Lugdunum.

Digges, Leonard, and Thomas Digges. 1576. *A Prognostication Everlasting*. London.

Dijksterhuis, E. J., ed. 1955. *The Principal Works of Simon Stevin*. Vol. 1, *General Introduction, Mechanics*. Amsterdam: Swets & Zeitlinder.

Drake, Stillman. 1958. "Galileo Gleanings III: A Kind Word for Sizzi." *Isis* 49(2): 155–65.

Euclid. 1956. *The Thirteen Books of the Elements*. 2nd ed. Translated by Thomas L. Heath. 3 vols. Mineola, NY: Dover Publications.

Fibber McGhee and Molly. 1949. *Dressmaker's Dummy*. NBC Radio, March 22 broadcast.

Finocchiaro, Maurice A. 1989. *The Galileo Affair: A Documentary History*. Berkeley: University of California Press.

———. 2010. *Defending Copernicus and Galileo: Critical Reasoning in the Two Affairs*. Dordrecht: Springer.

———. 2013. "Galileo under Fire and under Patronage." In *Ideas Under Fire: Historical Studies of Philosophy and Science in Adversity*, edited by Jonathan Lavery, Louis Groarke, and William Sweet, 123–44. Lanham, MD: Rowman & Littlefield.

———. 2014. *The Routledge Guidebook to Galileo's Dialogue*. New York: Routledge.

Fracastoro, Girolamo. 1574. "Homocentricorum, sive de stellis." In *Opera omnia: In unum proxime post illius mortem collecta, etc.* Venice.

Galilei, Galileo. 1663. *A Discourse Concerning the Natation of Bodies upon, and Submersion in, the Water*. Translated by Thomas Salusbury. London.

———. 1856. *Le opere di Galileo Galilei: Opere astronomiche. 1842–1853 prima edizione completa*. Florence: Società Editrice Fiorentina.

———. 1957. *Discoveries and Opinions of Galileo*. Translated by Stillman Drake. New York: Anchor Books.

————. 2001. *Dialogue Concerning the Two Chief World Systems: Ptolemaic and Copernican.* Translated by Stillman Drake. New York: Modern Library.

Graney, Christopher M. 2011. "Coriolis Effect, Two Centuries before Coriolis." *Physics Today* 64(8): 8–9.

————. 2013. "Stars as the Armies of God: Lansbergen's Incorporation of Tycho Brahe's Star-Size Argument into the Copernican Theory." *Journal for the History of Astronomy* 44: 165–72.

————. 2015. *Setting Aside All Authority: Giovanni Battista Riccioli and the Science against Copernicus in the Age of Galileo.* Notre Dame, IN: University of Notre Dame Press.

Graney, Christopher M., and Timothy P. Grayson. 2011. "On the Telescopic Disks of Stars—A Review and Analysis of Stellar Observations from the Early 17th through the Middle 19th Centuries." *Annals of Science* 68: 351–73.

Handy, Richard, Deirdre Kelleghan, Thomas McCague, Erika Rix, and Sally Russell. 2011. *Sketching the Moon: An Astronomical Artist's Guide.* New York: Springer.

Hart, Charles A. 1959. *Thomistic Metaphysics: An Inquiry into the Act of Existing.* Englewood Cliffs, NJ: Prentice-Hall.

Hawtrey, Stephen Thomas. 1874. *An Introduction to the Elements of Euclid.* London: Longmans, Green, and Co.

Heath, T. L. 1897. *The Works of Archimedes.* Cambridge: Cambridge University Press.

Heilbron, John. 2010. *Galileo.* Oxford: Oxford University Press.

Kelleghan, Deirdre, and Christopher M. Graney. 2016. *Astronomers Who Observe Visually and Draw What They See are Cool—and Valuable to the History of Astronomy! [plus comments].* September 7. http://www.vofoundation.org/blog/astronomers-observe-visually-draw-see-cool-valuable-history-astronomy/.

Kepler, Johannes. 1596. *Prodromus dissertationum cosmographicarum, continens mysterium cosmographicum etc. Addita est erudita Narratio Georgii Joachimi Khetici, de Libris Reuolutionum etc.* Tubingen.

————. 1604. *Ad vitellionem paralipomena, quibus astronomiae pars optica traditur.* Frankfurt.

————. 2000. *Optics: Paralipomena to Witelo & Optical Part of Astronomy.* Translated by William H. Donahue. Sante Fe, NM: Green Lion Press.

Knight, Charles, ed. 1868. *Arts and Sciences: Or, Fourth Division of "The English Encyclopedia."* Vol. 8. London: Bradbury, Evans, and Co.

Lattis, James M. 1994. *Between Copernicus and Galileo: Christoph Clavius and the Collapse of Ptolemaic Cosmology.* Chicago: University of Chicago Press.

Leopold of Austria. 1489. *Compilatio Leopoldi ducatus Austrie filij ad astrorum scientia decem continens tractatus.* Augsburg.

Locher, Johann Georg. 1614. *Disquisitiones mathematicae, de controversiis et novitatibus astronomici.* Ingolstadt.

Macrobius. 1952. *Commentary on the Dream of Scipio: Translated with an Intro-duction and Notes by William Harris Stahl.* Translated by William Harris Stahl. New York: Columbia University Press.

Macrobius, Ambrosius Theodosius. 1556. *In somnium Scipionis.* Lugdunum.

Marius, Simon. 1614. *Mundus Jovialis.* Nurnberg.

Marius, Simon, and A. O. Prickard. 1916. "The 'Mundus Jovialis' of Simon Marius." *The Observatory* 39(504–7): 367–81, 402–12, 443–52, 498–503.

Mayall, R. Newton, Margaret Mayall, Jerome Wyckoff, and John Polgreen. 1977. *The Sky Observer's Guide.* New York: Golden Press.

Monaco, Giuseppe. 1990. "Lorenzo Respighi and Star Scintillation." *Memorie della Società Astronomia Italiana* 61: 819–27.

Morin, Jean Baptiste. 1631. *Famosi et antiqui problematis de telluris motu, vel qui-ete, hactenus optata solutio.* Paris.

Nelli, Giovan Battista Clemente. 1793. *Vita e commercio letterario di Galileo Galilei: Nobile e patrizio fiorentino, mattematico e filosofo sopraordinario de Gran Duchi de Toscana Cosimo e Ferdinando II.* Vol. 1. Losanna.

North, Thomas. 1898. *Plutarch's Lives Englished.* Vol. 3. London: J. M. Dent and Co.

Peucer, Kaspar, and Konrad Dasypodius. 1568. *Hypotyposes orbium coelestium.* Argentoratum.

Piccolino, Marco, and Nicholas J. Wade. 2014. *Galileo's Visions: Piercing the Spheres of the Heavens by Eye and Mind.* Oxford: Oxford University Press.

Plato. 1961. *The Collected Dialogues of Plato, Including the Letters.* Edited by Edith Hamilton and Huntington Cairns. Princeton: Princeton University Press.

Possevinus, Anthonius. 1593. *Bibliothecae selectae pars secunda.* Rome.

Proclus. 1560. *Elementorum librum commentariorum.* Translated by Francis Baro-cius. Patavium.

———. 1792. *The Philosophical and Mathematical Commentaries of Proclus on the First Book of Euclid's Elements.* Translated by T. Taylor. Vol. 1. London.

Reeves, Eileen. 1997. *Painting the Heavens: Art and Science in the Age of Galileo.* Princeton: Princeton University Press.

Reinhold, Erasmus, and Georg Peurbach. 1580. *Theoricae novae planetarum, etc., Inserta item methodica tracatio de illuminatione Lunae.* Wittenberg.

Riccioli, Giovanni Battista. 1651. *Almagestum novum.* Vol. 1 (Tomus Primus). Bologna.

Risner, Friedrich. 1572. *Opticae thesaurus, Alhazeni Arabis libri septem, etc. Vitel-loni Thuringopoloni opticae libri X, etc.* Basel.

Rosen, Edward. 1959. *Three Copernican Treatises (including the* Narratio Prima *of* Rheticus*).* 2nd ed. New York: Dover Publications.

Scheiner, Christoph. 1612. *De maculis solaribus et stellis circa Jovem errantibus, ac-curatior cisquisitio.* Augsburg.

Scientific American. 1896. "Minute Workmanship." April 18: 250.

Stahl, William H., and Martianus Capella. 1977. *Martianus Capella and the Seven Liberal Arts.* Vol. 2, *The Marriage of Philology and Mercury.* New York: Columbia University Press.

Stevin, Simon. 1605. *Liber primus staticae, de staticae Elementis.* Vol. 2 (Tomus Secundus), in *Hypomnemata Mathematica.* Lugdunum Batavorum: Ioannis Patii.

Van Helden, Albert. 1985. *Measuring the Universe: Cosmic Dimensions from Aristarchus to Halley.* Chicago: University of Chicago Press.

Van Helden, Albert, and Eileen Reeves. 2010. *On Sunspots.* Chicago: University of Chicago Press.

Vitruvius. 1914. *The Ten Books on Architecture.* Translated by Morris Hickey Morgan. Cambridge, MA: Harvard University Press.

Wanley, Nathaniel. 1806. *The Wonders of the Little World: Or, a General History of Man, Displaying the Various Faculties, Capacities, Powers, and Defects of the Human Body and Mind.* Vol. 1. London.

Wesley, John. 1823. *A Survey of the Wisdom of God in the Creation: Or a Compendium of Natural Philosophy.* Vol. 2. New York: N. Bangs and T. Mason.

Westman, Robert S. 2011. *The Copernican Question: Prognostication, Skepticism, and Celestial Order.* Berkeley: University of California Press.

INDEX

air
    as cause of celestial phenomena, 50,
        75, 78–82, 100, 130n.171
    and circular motion of heavy things,
        33–35, 121n.86, 121n.88
    in the system of elements, 17, 40,
        49–50, 112n.54, 113 (fig.)
Alphonso X of Castile, 55–56
anti-Copernicans, characterization of,
    xi–xii, xv, xxiv
Apelles. *See* Scheiner, Christoph
Archimedes, 9–10, 22, 29–30
Archytas, 9
Aristarchus, 29–30
Aristotle, 7–9, 41, 90–91
    contradicted, 23
    various works of, 8
arithmetic, xiii, 10, 12, 17
    supports astronomy, xiii, 17
astrology, xiii, 11, 16
astronomy, 11–12
    defined, xiii, 16–17
    and the infinite, 17
atoms (elementary particles), xxiii,
    17–18, 23

Brahe, Tycho
    appropriation of system from
        Capella, 59–60
    and Coriolis effect, 121n.92
    data from, 28–29, 99, 134n.203
    and star sizes, xxi–xxii, 30, 117n.75
    system of, xii, 58–59

Capella, Martianus, 57
    system of, 59–60
Cardano, Girolamo, 69
Cicero, 9
Clavius, Christopher, 11, 55–58, 60, 90
    and telescopic discoveries, 57
colors, 46–47, 86–87
    of celestial bodies, 60–61, 83
Copernicus, Nicolaus
    arguments against, xviii–xxiii, 27–39,
        105 (*see also* falling-bodies
        argument; star-size argument)
    astronomers following, xvi, 25
    On the Revolutions, xi, 27, 29, 33, 36
    and parallax, xx
    and the size of the universe, 29,
        117nn.70–71, 120n.82

CHRISTOPHER M. GRANEY

is professor of physics and astromony at

Jefferson Community & Technical College in Louisville, Kentucky.

He is the author of *Setting Aside All Authority:*

Giovanni Battista Riccioli and the Science against Copernicus in

the Age of Galileo (University of Notre Dame Press, 2015).

CPSIA information can be obtained
at www.ICGtesting.com
Printed in the USA
BVOW08s2317051017
496808BV00003B/9/P